Science and the Good

FOUNDATIONAL QUESTIONS IN SCIENCE

At its deepest level, science becomes nearly indistinguishable from philosophy. The most fundamental scientific questions address the ultimate nature of the world. Foundational Questions in Science, jointly published by Templeton Press and Yale University Press, invites prominent scientists to ask these questions, describe our current best approaches to the answers, and tell us where such answers may lead: the new realities they point to and the further questions they compel us to ask. Intended for interested lay readers, students, and young scientists, these short volumes show how science approaches the mysteries of the world around us and offer readers a chance to explore the implications at the profoundest and most exciting levels.

Science and the Good

The Tragic Quest for the Foundations of Morality

James Davison Hunter
Paul Nedelisky

Yale UNIVERSITY PRESS
NEW HAVEN AND LONDON

Templeton Press

Yale University Press books may be purchased in quantity for educational,
business, or promotional use. For information, please e-mail sales.press@yale
.edu (U.S. office) or sales@yaleup.co.uk (U.K. office).

Designed and set in Hoefler Text by Gopa & Ted2, Inc.

Printed in the United States of America.

Library of Congress Control Number: 2018947847
ISBN 978-0-300-19628-3 (hardcover : alk. paper)

A catalogue record for this book is available from the British Library.

This paper meets the requirements of ANSI/NISO Z39.48-1992
(Permanence of Paper).

10 9 8 7 6 5 4 3 2 1

Contents

CONTENTS

Acknowledgments

WE WOULD like to thank the participants of a workshop, sponsored by the Institute for Advanced Studies in Culture, that was devoted to discussion of the first full draft of this book. We are deeply indebted to Tal Brewer, Jeff Guhin, Charles Mathewes, James Mumford, Trevor Quirk, Robert Reed, Rebecca Stangl, and Jay Tolson. The close reading and ranging commentary they provided was deeply critical, uniformly insightful, and enormously encouraging. Along the way, we benefited from the conversation and advice we received from several others. David Smith, Lauren Turek, Ethan Podell, Nick Frank, Nicholas Olson, and Andrew Walker, in addition to the audience of the University of Virginia Blandy Experimental Farm Philosophy Retreat, all provided helpful criticism of the very early drafts and some of the completed chapters of the manuscript. To them we are very grateful. We would also like to extend our deep appreciation to Andrew Morgan, Jonathan Barker, Emily Hunt Hinojosa, and John Nolan for providing invaluable research assistance in the task of mapping the cartography of philosophical history, academic ethics, and moral science.

We are also immensely grateful to our editors, Susan Arellano and Bill Frucht. One could not ask for wiser or more insightful counselors.

No work of scholarship happens outside of a community of discourse. For the rich intellectual community of the Institute for Advanced Studies in Culture—for our colleagues (in addition to those mentioned above) Nick Wolterstorff, Ryan Olson, Joe Davis, Matt Crawford, Tony Tian-Ren Lin, Rachel Wahl, Murray Milner, Garnette Cadogan, Stephen White, and Josh Yates and the logistical and financial support the Institute provided— our profound gratitude.

Preface
THE ARGUMENT, IN BRIEF

W HO COULD DENY IT? Modern science since the Enlightenment has been nothing short of a wonder. Its achievements in solving enduring riddles over the past half-millennium have been astonishing. Put aside for a moment some of the malevolent ways science has been used—the method itself, within many spheres of inquiry, has generated a range of new knowledge and insight that is nothing if not breathtaking.

For most of us, the intricacies of scientific knowledge are unfathomable, endlessly so. It is a realm far out of reach of the understanding of most mortals, which is why science can be, ironically, so mystifying. Much of the authority of science in the modern and now late modern world derives from both its extraordinary track record and from the esotery of it all—a knowledge possessed, albeit in fragments, almost exclusively by those rare individuals credentialed with advanced degrees in particular scientific fields.

Is it any wonder that we would give science, and those who

speak for it, the benefit of the doubt? But even for a subject as important as human morality?

The possibility is arresting, to say the least.

Indeed, it is quite a bracing experience to go into a bookstore or browse online and see titles claiming to show "how science can determine human values," to uncover the "science of moral dilemmas," to disclose "the biological basis of morality" or "the science of right and wrong," to reveal "the universal moral instincts caused by evolution," to explain how a certain molecule is "the source of love and prosperity," to describe "how nature designed our universal sense of right and wrong," or to demonstrate "what neuroscience tells us about morality."[1] These claims are all taken from the titles of recent books and articles, and these claims are pervasive.

What pluck! These titles would seem to defy the age-old rule called "Hume's Law" that you can never derive an "ought" from an "is": that there is a decisive boundary separating prescription from description.

Could it be that this is no longer true? Have new technologies and new ways of thinking rendered the rule obsolete?

The very idea that scientists or philosophers of science could reach into the physical universe and demystify the nature of "the good" to reveal the physiological and evolutionary mechanisms behind ethics or the neurochemistry underneath morality is exciting, and in some ways intimidating, even if the idea is not fully persuasive.

The purpose of this book is to examine those claims. They are rooted in a longstanding and impassioned quest to find a scientific foundation for morality. When did this quest originate and why? How has it evolved? What is its current standing? What has it accomplished, and where is the quest leading us? What is at stake is not merely academic. Rather, what is at stake is the yearning to coherently address some of the knottiest problems of the modern world—not least, *how and upon what foundation do we build a good and just society?*

THE ARGUMENT, IN BRIEF

The heart of our argument is found in the story of this four-hundred-year quest to establish a science of morality. The story begins at the advent of the modern West, in a time when Europe was riven with conflict over the right, just, and moral ordering of society. Traditional religious beliefs and medieval philosophy had not only conspicuously and tragically failed to bring order and peace to an increasingly pluralistic world but had made such hopes ever more elusive. Against these failures, the emergence of science promised a new way forward in all spheres of life. After all, science had achieved extraordinary success in understanding the natural world and in addressing a range of human problems. Why couldn't it also solve enduring moral problems, not least of which was the puzzle of how to fashion a good and peaceable society? This is important, for what is at stake in this question was nothing less than the possibility of a new foundation for human flourishing.

Some of brightest minds of the Enlightenment looked to science to address these persistent questions. Over the next few centuries, the quest followed several paths. Some thought that moral reality could be established experimentally by observing which human laws promoted peace and concord. Some thought a mechanical theory of the mind would reveal everything we need to know about the moral realm. Others looked to the measurement of human pleasure to define morality. Still others thought that the dynamics of human evolution produced morality as a tool of survival.

But after four hundred years, the ideal of understanding moral reality scientifically through observation and demonstration—in the way that truths in astronomy and medicine were understood—continued to confound. The various paths to ground morality in science seemed to end—in part because none had succeeded, and in part because science fragmented into specialized disciplines, none of which focused on morality. By the end of the nineteenth century, the prospects of establishing a scientific foundation for morality were not at all hopeful.

In the 1970s, however, with the reintegration of multiple scientific disciplines along with several of the older, more philosophical paths, the quest reemerged with renewed vigor. A new synthesis, aided by advancing technology, had created new enthusiasm for fulfilling this time-honored quest.

But has the new moral science actually brought us closer to achieving its aspirations?

Sadly, no. What it has actually produced is a modest though interesting descriptive science of moral thought and behavior. We now know more, to take one example, about what is happening at the neural level during moral decision-making.

Yet many of its proponents claim much more for these types of findings than the science can justify. While some of this over-reaching is due to honest mistakes or misunderstandings about what science has shown, some of it appears fraudulent, designed to capitalize on science's prestige and the public interest in practical moral advice. In the end, the new moral science still tells us nothing about what moral conclusions we *should* draw.

This is not happenstance. There are good reasons why science has not given us moral answers. The history of these attempts, along with careful reflection on the nature of moral concepts, suggests that empirically detectable moral concepts must leave out too much of what morality really is, and moral concepts that capture the real phenomena aren't empirically detectable. Whether they realize it or not, today's practitioners of moral science face this quandary, too.

But here the story takes a surprising turn. While the new science of morality presses onward, the idea of morality—as a mind-independent reality—has lost plausibility for the new moral scientists. They no longer believe such a thing exists. Thus, when they say they are investigating morality scientifically, they *now* mean something different by "morality" from what most people in the past have meant by it and what most people today still mean by it. In place of moral goodness, they substitute the merely useful, which is something science *can* discover. Despite using the language of morality, they embrace a view that, in its net effect, amounts to moral nihilism.

When it began, the quest for a moral science sought to discover the good. The new moral science has abandoned that quest and now, at best, tells us how to get what we want. With this turn, the new moral science, for all its recent fanfare, has produced a world picture that simply cannot bear the weight of the wide-ranging moral burdens of our time.

PART I
Introduction

CHAPTER 1

Our Promethean Longing

C AN SCIENCE be the foundation of morality?
The social implications of this question are enormous.
We live in a time rife with disagreement, conflict, and violence—clashes that are almost always rooted in competing conceptions of the good. How, then, do we resolve such disagreements? Is there a way to arbitrate these disputes? Surely in a day of cosmopolitan sophistication, there must be some way to mediate them—some compelling logic that could provide a common foundation for moral belief and commitment.

There are many for whom this question is absurd on the face of it—who say there is a prima facie case against science ever being the foundation for morality. Their quick dismissal ignores the fact that some of the brightest minds in science and philosophy are confident that science *can* be the foundation of morality. Indeed, public discourse is awash with books that claim this very thing. All of this suggests that we may be at the start of a new age in which science provides clarity and insight into vexing moral questions.

The skeptics' quick dismissal also ignores how central—and passionate—the quest for such a foundation has been to Western thought over the last four hundred years. Over centuries, the pursuit has been nothing if not ardent.

The question, then, matters, and it matters a great deal. But why?

The Dilemma of Difference

When one looks at this history carefully, one can see that, from the beginning, the animating force behind the quest for a scientific foundation for morality has been the desire to address the problems of moral difference and complexity and, more to the point, the conflict and confusion they generate. Many, to be sure, are also motivated by a pure search for truth. But even the search for truth is always embedded in a time and place and is strongly influenced by the contingencies of history and culture. Those contingencies always point to the overriding concern with the problem of difference.

The quest maps roughly onto the story of modernity. That story is, among other things, a story of the shrinking of the world in ways that bring in closer proximity different cultures and different ways of life. While the plurality of cultural difference has always existed, the past half millennia has amplified that development in ways that previous generations could not have imagined. The problem is that the coexistence of cultures is always accompanied by competing claims on shared public space, contradictory interests, and the inequities of power and privilege. Precisely because difference nearly always plays out at fundamental levels of human belief, and because the conflicts matter so concretely in human experience, they are nearly always accompanied by suspicion, tension, the suppression of legitimate claims and interests, latent antagonism, and sometimes open conflict and violence. The accumulated costs of these differences are beyond comprehension.

In the early decades of the twenty-first century, those differences are intensified to the point that we would say that the now globally interconnected world is constituted by these deep social, cultural, and political differences. As is plain to see, these differences are anything but abstractions but rather continue to bear on issues fundamental to the well-being of all human beings— order, security, freedom, fairness, health, and wholeness. Is there an issue of public policy or foreign policy that is not morally fraught? Immigration, health care, racial inequality, care for the elderly and for the poor, education, aid to victims of natural disaster, international trade, and war are all laced with difficult moral questions that have no easy answers and that more often than not lead us to fundamental disagreements over what is right and wrong, good and evil, just and unjust. And underneath the many specific questions are more fundamental disagreements about what constitutes the good life, the good society, and the good world.

No one's motives are entirely pure. All of us operate with, at best, mixed and conflicting intentions. Yet most antagonists have been and are sincere in their desire for human flourishing, at least on their terms. Whatever else may motivate them, they also happen to disagree fundamentally and mostly sincerely on what is true, right, and good. Here, too, the costs of these disagreements are often beyond reckoning.

THE PROBLEM OF COMPLEXITY

The dilemma of difference is only made more confounding by the sheer complexity of the modern and now late modern world.

The explosion of knowledge that came with modernity is difficult to fully comprehend. Between 1517 and 1550, approximately
150,000 new books were published in Europe. This was at least
four times as many as had appeared during the entire fifteenth
century. Between 1517 and 1523 alone, one could find 400 print-
ers, 125 places of publication, and approximately 900 authors.[1] At
that time, this represented an extraordinary growth in the world's
knowledge. Five hundred years later, of course, the growth in
information and knowledge has surged exponentially.

We now live in an age of information superabundance. It is often
noted that more information has been produced in the last thirty
years than in the previous five thousand. Around 1,000 books are
published internationally every day, and the total of all printed
knowledge doubles every five years.[2] Yet printed documents only
make up .003 percent of total information. The Internet and other
digital technologies, of course, have only intensified the production, collection, and distribution of information. The world has
produced 300 exabytes (300,000,000,000,000,000,000,000 pieces)
of information—and produces between 1 and 2 exabytes[3] of
unique information per year, which is roughly 250 megabytes for
every man, woman, and child on earth. To make this a little more
concrete, 300 billion emails, 200 million Tweets, and 2.5 billion
text messages course through our digital networks every day.[4]
Add to this the 85,000 hours of original programming produced
every day by over 21,000 television stations and the 6,000 hours
of YouTube video produced every hour.[5] The weekday edition of
the *New York Times* contains more information than the average
person in seventeenth-century England was likely to come across
in a lifetime.[6]

We are overcome by a tsunami of information. Is there clarity,
wisdom, or truth to be had in the midst of this complexity? If so,

how do we sort through it all? The puzzles posed by difference and complexity are built into the modern world. Given the conflict, disorder, confusion, and human suffering that follow in the wake of our deepest differences, and given the massive complexity of modern knowledge and information, questions arise: What is Justice? Fairness? Equity? How do we live together at peace with our deepest moral differences? And if we can't agree on shared principles or ideals and their application, on what grounds do we adjudicate our disagreements?

THE PROMISE OF SCIENCE

Some see science as the only method that offers any hope of being such a rational arbiter. After all, the methods of science—observation, experimentation, theory building—have delivered a persuasive picture of the physical universe. This has brought a consensus in the physical sciences that stands in stark contrast to the disorderly tumult of moral opinion. Iranian scientists, for example, accept and employ the same view of physics as do scientists in France; Chinese scientists and Norwegian ones operate with the same understanding of chemistry. Yet the moral viewpoints across these cultures differ in the extreme.

And so we arrive at this bright thought: perhaps science can do for morality what it has done for physics, chemistry, biology, astronomy, and mathematics, and the technologies that are based upon them. This is the question that animates this book. Can the methods of science provide rational and compelling answers to questions of right and wrong, good and bad, and how we ought to live? Can science be the foundation of morality?

WHAT IS AT STAKE

There is, then, on the surface, a bland, inoffensive, even somnolent academic quality to this question. But the question behind the question is, can science rise above our differences, cut through the complexity, and serve as the foundation of a just and humane social order? Can there be a "science of human flourishing"?[7] This is the enduring question at stake in the quest to find a scientific foundation for morality.

Bertrand Russell once reflected on the possibility of a "scientific society," saying that

> While upheavals and suffering have hitherto been the lot of man, we can now see, however dimly and uncertainly, a possible future culmination in which poverty and war will have been overcome, and fear, where it survives, will have become pathological. The road, I fear, is long, but that is no reason for losing sight of the ultimate hope.[8]

If science can demonstrate a foundation for morality, then there is the potential that confusion will lessen and conflict will subside. For if there is a true science of morality, then the good life can be found and demonstrated to be so, thereby settling disagreements just as disagreement about the composition of water was settled by the demonstrations of chemistry. We will know the nature of the good, how to live a good life, and how to build a just and peaceable social and political order. Empirical investigation will generate a *pax scientia*.[9] As one enthusiast put it, "Only a rational understanding of human well-being will allow billions of us to coexist peacefully, converging on the same social, political, economic, and environmental goals."[10] This is the Promethean longing that animates this question and the debate surrounding it.

We want to be clear. Our question is not, "Can science teach us anything about morality?" On the face of it, that is a hard question but a fair one, and rather uncontroversial. Anyone with intellectual curiosity might be interested in this—and why not? On the one hand, you have morality—the sum and substance of the good in human behavior and society—which is found (differently) in every human civilization, and yet its nature and workings remain a great and unsolved puzzle. On the other hand, you have science, a method of rational inquiry that systematically builds knowledge from testable explanations. Clearly science's centuries-long record of accomplishment in understanding the conundrums of the universe is beyond comparison. So why not morality? Perhaps science can help unravel this riddle as well.[11]

Those who argue that science is or should be the foundation for morality are generally making an epistemological claim about the superiority of science over other forms of knowledge. Debate about this claim is almost as intense as the disagreement it is supposed to resolve. Why? What is at stake here is the viability of a certain comprehensive view of reality called *naturalism*. Naturalism is the idea that, at bottom, everything that exists can be understood in the terms used by science. So, of course, naturalists tend to see science as the primary, best, or only way to know things.[12] Naturalism is in competition with perspectives that look to other, often *non*scientific and *non*empirical bases for truth, knowledge, understanding, and wisdom. Among these nonscientific bases are intuition, common sense, introspection, various traditions, religion, and pure reason.

At this point, the relationship between science and morality becomes a critical, if not central, battle in the larger culture war. This larger conflict was never merely about a handful of unconnected issues like abortion, homosexuality, public funding for the arts, the relationship between church and state, and the like, but rather about the animating visions of the "good society" and the moral authority upon which these competing visions are based. These questions touch on what people most cherish, which helps to explain why passions become so inflamed. Few people will easily or willingly compromise on what is sacred to them, and it is precisely for this reason that attempts at persuasion have little effect. As in other areas of cultural conflict, *ir*rationality, dogma, and fanaticism can be found on all sides, not least among those who claim the authority of autonomous reason and scientific impartiality. The question about the relationship of science to morality goes right to the heart of these tensions.

Importantly, it is not just competing ideas of truth that are in play. There are also powerful interests at stake, for how these questions are answered will say much about the allocation of power and privilege.

POINTS OF CLARIFICATION

The Question

So just to be clear, we are not asking the question, "Can science tell us anything about morality?" Surely it can, especially about the descriptive aspects of morality—what people think about morality and what physical processes underlie moral thought and behavior.

Neither are we asking, "Can ethical naturalists—those who hold that good and bad or right or wrong are part of a purely natural world—be moral?" On the face of it, the question is absurd.

To be human is to be an active agent within a moral universe, and just like people of religious faith, ethical naturalists are capable of both the most noble and the most despicable acts imaginable.

Rather, our focus and main question is whether science can do for morality what it does for chemistry and physics—resolve differences with empirical evidence. In short, "Can science demonstrate what morality is and how we should live?"

To challenge that project is in no way to challenge the validity of science itself. To repeat: nothing we argue here poses any threat to the value or validity of science itself, or to the naturalism its methods presuppose. After all, even if we are correct that science can't tell us about some things, it doesn't follow that science can't tell us anything at all. Compare: a metal detector cannot tell you everything about what's buried at the beach, but it can tell you about the buried metal things. Similarly, science may not be able to tell us how to live, but it can tell us about physical reality and its laws.

Neither would it follow that there are no longer any rules for rational inquiry. Even if science cannot tell us what we want to know about morality, there are other means of assessment: our ordinary experience, our nonempirical yet introspective awareness, our understanding of human motivations, our basic rational facility for understanding and abstracting the essential features of things, and theory-building with the data obtained from these sources. To be sure, none of these can be taken as a faultless source of truth. Instead, just as in the practice of science, each must be examined, compared to the broader evidence, and reevaluated in light of new experiences. Nevertheless, this is a far cry from mere fantastical speculation. The metal detector analogy is again helpful: just because there are things buried that your metal detector cannot tell you about, it doesn't follow that just anything goes in identifying the other buried objects. It's still reasonable to use shovels to find other buried items, and it still isn't reasonable to use Ouija boards.

Our Approach

The subject of this study is the *discourse* surrounding the relationship between science and morality. This discourse is certainly an academic discourse, which is why it must be understood academically, but it does not fall neatly within any particular academic field: it is not psychology, biology, neurochemistry, or any of the sciences per se, nor is it ethics per se, nor philosophy or history. It exists on the peripheries of all of these fields—marginal at best to the mainstream of these academic disciplines, especially ethics and philosophy. Rather, the discourse we seek to understand is a *hybrid* whose contributors are philosophers, ethicists, psychologists, biologists, neurologists, physicists, and so on.

Because the discourse itself is interdisciplinary, it would be entirely inappropriate to approach it from a narrowly disciplinary perspective. Our own backgrounds, in the historical sociology of knowledge and culture (Hunter) and in philosophy (Nedelisky), give us as much of a claim to engage this subject as practitioners in any other field. Our particular areas of expertise are not within the cluster of fields that dominate this discourse—psychology, biology, neurochemistry, and ethics. But as will become clear, we do not attempt to innovate within these fields, and our arguments seldom require technical expertise. Our task is to understand, contextualize, interpret, and engage this body of knowledge from the vantage point of our home disciplines. In this respect, we regard our position as outsiders to this club as a distinct advantage—for two reasons.

First, the subject is far too important to be relegated to specialists. It is consequential for everyone, and to close it off from broader public (and scholarly) discussion is the surest way to achieve academic insularity, and thus to guarantee self-confirming assessments of validity and significance. We need to open the windows and let the fresh air of broad intellectual inquiry blow through.

Second, one of our central objectives in this book is to bring a perspective that only those outside the particular academic circle that dominates in this discourse can bring, and to do so in a way that is accessible to a broader public.

Our Method

What do we mean when we say that, in its totality, the literature, debate, and discussion promoting a science of morality—especially today's "new moral science"—*has come to constitute a discourse?*

Put simply, we mean that the conversation about the science of morality has taken on a life of its own that goes beyond any particular academic discipline. The questions asked and answers proposed may draw on any number of fields, but ultimately these questions and answers are what they are, apart from the narrow focus of neuroscience, academic ethics, or evolutionary biology. This discourse is not just a body of evolving knowledge but a mood, a disposition, a set of affinities, an arrangement of rules, and a collection of institutional and symbolic resources; hence, it is its own culture, with a distinctive life of its own.[13]

In this respect, it is not unlike the discourse that constitutes, say, conservatism or liberalism or identity politics or Christian fundamentalism. In this case, the discourse and the discursive community that generates and propagates the new moral science, while certainly made up of moral scientists, philosophers, and polemicists, is not reducible to any one of them. Some contribute a little and others contribute a lot. Some are closely aligned with the leading ideas of the discourse, and others distance themselves from it on this or that point. Among those who give voice to this discourse, some are exceedingly careful and circumspect while others tend toward bluff and bluster. About any discourse, one can always say, "I can tell you five ways that this person or that person doesn't fit." Such a view misconstrues the nature of culture and, in particular, the discourse that constitutes the new moral science.

How, then, does this discourse take form and find expression?

Like any discourse, it is constituted by its own distinct and complex cultural economy. At its heart are networks of scholars and public intellectuals, more often than not, attached to colleges and universities, and the work they produce. Today, these scholars typically find a home in graduate faculties and programs in psychology and philosophy with concentrations in moral psychology. These are often organized in college or university labs, of which there are dozens, including the Moral Psychology Research Lab at Harvard; the Morality Lab at Boston College; the Values, Ideology, and Morality Lab at the University of Southern California; and the Social and Moral Cognition Lab at Columbia.[14] These are similar to grant-funded projects such as the New Science of Virtues project at the University of Chicago[15] and the Cambridge Moral Psychology Research Group.[16] These departments, laboratories, and programs connect via pan-university networks such as the Moral Psychology Research Group,[17] the Moral Research Lab,[18] and Yourmorals.org. Out of these programs and networks comes extraordinary scholarly output, much of it published in strictly academic journals such as *Nature, Science, Cognition, Brain, Journal of Personality and Social Psychology, Cognition and Emotion, Journal of Neuroscience, Cognitive Science, Journal of Research in Personality, Emotion, Journal of Applied Social Psychology, Ethics, Social*

Neuroscience, and *Journal of Moral Education*, as well as more popular journals such as *New Scientist*, *Scientific American*, and *Discovery*. The subject is covered in the mainstream press as well, including prestige news media like the *New York Times* and *New York Times Magazine*, the *Washington Post*, *Slate*, and National Public Radio. Then, of course, there are immense numbers of academic and popular books making larger arguments about the new moral science. Among the most visible authors have been Frans de Waal, Patricia Churchland, Jonathan Haidt, Joshua Greene, Steven Pinker, Marc Hauser, Owen Flanagan, Paul Thagard, Alex Rosenberg, Michael Ruse, Sam Harris, Paul Zak, and Michael Shermer. Not least, the blogosphere is full of discussion about the new moral science.[19] The discourse is also carried by innumerable conferences. The vast majority are academic, though some bring together public intellectuals, such as the important Edge conference on "The New Science of Morality" in 2010.[20] Providing a much greater megaphone are platforms that address the public, including many TED conferences and the Aspen Institute's Ideas Festival.[21]

Financial support for this work comes from sources like the Edge Foundation, the Nour Foundation, and the National Science Foundation. By far the most generous funder of the new moral science has been the Templeton Foundation. Founded by the late billionaire Sir John Templeton, it exists to carry forward its founder's mission to discover new "spiritual information" through science.[22] It has done this by injecting hundreds of millions of dollars into a variety of research programs, many of which could be classified as scientific approaches to morality, including millions to the psychological study of human flourishing and the science of character and character development.

But while the Templeton Foundation may be the biggest

funder, arguably the most strategic foundation for cultivating the broader discourse within which the discourse on science and morality finds a platform is the Edge Foundation, founded and led by John Brockman.[23] With its stated mission of reframing public culture broadly in light of science, and with nearly nine hundred scientists, philosophers, and public intellectuals contributing to its website and various publications, the Edge Foundation is the most important organization promoting the plausibility of a new science of morality to the general public.

Why does this discourse matter?

While this discourse may be marginal to mainstream academic philosophy and ethics, it is thriving on the margins. It has generated and sustained many careers inside the academy, yet its greatest impact is outside the disciplinary guilds. As it is popularized, it assumes a disproportionate influence in shaping public opinion and discussion. Its concepts, assumptions, history, self-understanding, propositions, and inferences underwrite much of the billion-dollar industry that is positive psychology and the happiness movement—which in turn influences education, business, the military, and other institutions.

The idiom and sensibilities of the new moral science are also the idiom and sensibilities of the managerial elite of the dominant technocratic regime. It is especially influential with the middle strata of policy administrators, who desperately need a common language with which to speak authoritatively across so many differences.[24] Yet we also see it in other spheres in the reduction of performance, effectiveness, efficiency, and significance to the idiom of quantification, whether in business (metrics of performance for individuals, divisions, corporations), in education (metrics of literacy and numeracy, graduation rates, college acceptance, etc.), in medicine and health care (metrics of efficiency), in higher education (rankings of colleges and universities, rankings of departments and programs in fields, rankings of scholarly output and influence, etc.), in philanthropy (metrics of performance and influence), and so on. The new moral science is of a fabric with this widespread ethos, the regime upon which it is based, and the elites who depend upon it. Yes, the new moral science is marginal to the mainstream of philosophy, but it provides an intellectual legitimation for the unspoken technocratic hope found everywhere in contemporary public culture.

This is this sense in which the new moral science constitutes a discourse. The actual content of this discourse is certainly complex and varied, drawing on many academic sources—not least philosophical and ethical—and containing a wide range of disagreements and debate. Yet while it is far from monolithic, the discourse is also marked by a shared stock of evolving knowledge (e.g., what neuroscience has allegedly shown), a distinct cluster of dispositions and affinities (e.g., the tendency to understand the structure and function of the brain as a computer), and a loose set of rules and norms generated and disseminated within a cluster of different institutional structures. This is the evolving discourse we seek to understand.

In every discourse, there is a range of actors—from serious scholars to unabashed polemicists, from the few who influence public discussion to many who are invisible. In this book, we focus on those we consider the most serious and influential contributors. We want to represent and understand the new moral science in its most compelling form.

A Technical Matter (for Specialists)

Perhaps the best-known objection to a science of morality is the "Is/Ought Problem," also known as "Hume's Law."[25] As David

Hume explained back in the eighteenth century, people sometimes try to infer something about *what one ought to do* from claims about *what is* the case. But these are two different kinds of claims, and inferring an "ought" claim from an "is" claim requires an explanation to justify it. But people seldom give a justifying explanation. Hume argues there can be none.[26]

In general terms, the Is/Ought Problem should be a barrier to a science of morality because science studies what is, not what ought to be. It is concerned with factual observations of the world, not value judgments about ideals, goods, or duties. If it is impossible to get an ought from an is, and science can only study what is, then science cannot show us what we ought to do. This is the usual thumbnail sketch of the objection.

On the face of it, this looks like a formidable objection, yet as we will see, the history of the quest demonstrates that Hume's Law is far from universally accepted. The discourse that seeks to discover a scientific foundation for morality continues unabated, Hume's Law be damned.[27]

We believe the failure of Hume's Law to persuade is partly driven by historical and cultural circumstances: we *need* a science of morality to help us make sense of and navigate the often-chaotic difficulties of the modern world. Yet it is also driven by technical distinctions within philosophy itself. We will not wade into these philosophical technicalities at any great length. (Those who are interested can read the footnotes.[28]) Rather, we concentrate on the relationship between the legitimacy of the moral concepts being used and the degree to which they can be empirically investigated. This is, after all, what would make a science of morality especially interesting or promising: being able to empirically settle what we ought to do.

Our Agenda

This book should be read as both text and subtext. The text is about the relationship between science and morality. The subtext is an affirmative genealogy about a yearning to address some of the most difficult and challenging problems of the contemporary world—the challenge of pluralism, of confusion, and of the need for a common language for understanding the world and a common foundation for building a good and decent and just society.

As we say, these questions are too important to leave to scientists and scientifically minded ethicists alone. Members of academic guilds are often too close to the technical minutia of the debates of their field and too tied to the rules of the guild to see the wider significance of the subject in which they are experts. The fact is, these questions affect everyone, so everyone has a stake in the issue, not just those with PhDs in these areas.

We aim, then, for several things in this book. First and foremost, we want to understand, as history, the effort to give morality a scientific foundation. For all of the recent fanfare and the heated debate surrounding it, the question of whether science can provide a footing for moral questions is anything but new. This quest is one of the central strands in the story of the Enlightenment and of Western modernity. The current discussions on the subject are informed by centuries of audacious and innovative effort. This quest must be taken seriously on its own terms.

But it is also a quest that must be seen in its larger historical, cultural, and sociological milieu. Elements of the story have been told before, and very well, but we know of no account that attempts to give a narrative spanning the entire history. This narrative is crucial for several reasons. It explains the original motivations for the quest to find a scientific foundation for morality and shows how those motivations persist to the present day. The history of those aspirations and their repeated failures lets us see the prospects for this sort of science in a broader, clarifying perspective. Finally, understanding the longstanding motivations of the quest makes it easier to recognize the magnitude of the departure taken by present-day moral scientists. Though they use the traditional language of morality, they are doing something fundamentally different with it. The history of the quest brings this into relief. To this end, we bring together the methods and perspectives of intellectual history, philosophy, and the sociology of knowledge to tell a story of the roots, transformations, cultural logics, and unintended consequences of this historical quest. In this way, the quest to find a scientific foundation for morality is a window into the nature of the modern project and its fate.

We also sketch out what science has taught us about morality and the inherent challenges that present limitations to what can be known scientifically. We consider the philosophical and scientific adequacy of these efforts in light of the standards to which advocates themselves aspire.

Often, we find, the rhetoric gets well ahead of the science. The propensity to overreach is common within the academic scholarship, as is the tendency in the larger discourse to blur the boundaries between scientific description and moral prescription.

Within the science itself, of the many challenges that emerge, two are particularly daunting: the problems of definition and demonstration. Can we define morality in ways that the scientific and scholarly community can agree on, and that are adequate to the reality people experience? And once defined, can morality be empirically measured—its presence, its salience, its strength—in ways that are convincing? As we will see, these challenges operate at cross-purposes to each other. A kind of uncertainty principle seems to apply: the closer we get to a satisfactory definition, the further we get from empirical demonstration, and vice versa.

Our examination of the current state of this quest also leads to a perhaps surprising revelation: many of those pursuing a science of morality today are actually engaged in a very different project from what their historical predecessors were up to. Today's moral scientists no longer look to science to discover moral truths, for they believe there is nothing there to discover. As they see it, there are no such things as prescriptive moral or ethical norms; there are no moral "oughts" or obligations; there is no ethical good, bad, or objective value of any kind. Their view is, ironically—in its net effect—a kind of moral or ethical nihilism.

But here we have a puzzle. For one, few of the new moral scientists would use the term "moral nihilism" to describe their position, and fewer still would describe themselves as moral nihilists. What is more, the new moral science habitually uses the language of morality and moral prescription. How do we explain this seeming contradiction?

The resolution is found in the cultural logic they follow. As they would have it, even if there isn't anything we objectively "ought" to be doing, we still have to decide, on some basis, how to live and what to do. Without any real ethical standards, we look to social objectives as guides. The project, then, is about how science and technology can help us achieve these social goals. The role of science is to reveal how our moral psychology and neurochemistry work—or can be put to work—toward achieving those goals.

The problem is that these social objectives are, in the end, morally arbitrary, reflecting either fluctuating social tastes or the whims of those in power. In the end, as these thinkers see it, the "good" is a social engineering project, the foundation of which is an unmitigated, though rarely acknowledged metaphysical skepticism. This leaves the new moral science in a place where it is incapable of either critiquing the distortions of power and privilege or affirming higher moral ends that draw us to the possibilities of greater human flourishing—for everyone, but especially those without power and privilege.

From our vantage point we fully recognize the problems that attend a facile moral realism, the idea that moral reality exists in human life independent of experience, history, and social circumstance. That said, we are convinced of the irreducible normativity of human experience, and that out of that normativity it may be possible to discover enduring, even if not universally held, moral truths.

A FINAL QUESTION

It is hardly surprising to learn that tension and disagreement permeate highly academic philosophical discussions. Yet given all that is at stake here, the quarrels spill out into public discourse in ways that expose deep social and political divisions. For all of the difficult philosophical abstraction intrinsic to it, the quest to find a scientific foundation for morality unfolds within the vascular intricacies of contemporary social life. It is of a fabric with the history and culture that surround us. The final question, then, is: What does the quest reveal about our own time? What might it portend of our future?

The question "Is there a scientific foundation to morality?" has generated endless discussion and debate. But by attending to the difficulties of this very old question and to the longings that animate it, we may yet achieve a little more clarity about this confusing world we live in, even if we don't find a great deal more certainty.

PART II
The Historical Quest

CHAPTER 2

Early Formulations

I DEAS DON'T APPEAR out of thin air. They neither surface in discourse nor fade from collective memory of their own accord, but arise, take shape, and find expression under specific social and historical conditions. Ideas are always situated in society and history in ways that make them more or less plausible, more or less persuasive. This is clearly the case for the idea that science could be a foundation for morality.

In the West, the quest to find a scientific footing for morality begins in the fifteenth century. It is a rich and complex story with a compelling backdrop. There are innumerable actors and absorbing subplots in this history, and, while it is beyond our scope to provide an intricate account of this complexity, it is important to highlight certain key moments and key figures that mark the larger arc of this narrative. The reason is simple: the circumstances that gave rise to this Promethean longing are in some ways still with us. The impetus that gave rise to scientific attempts to ground morality keeps resurfacing in important ways—as do the arguments.

Early modern Europe was a place of profound social and intellectual transformation, as longstanding medieval worldviews and authority structures began to break apart in the face of new challenges. These challenges included: (1) the inability of old ways of knowing—philosophy, religious authority—to resolve exploding moral and political conflict; (2) a need for a convincing basis for shared international trade laws as global commerce swelled and broadened; (3) a sense that the world was bigger and more complex—in terms of natural, cultural, and moral phenomena—than older medieval conceptions could account for.

To many, situated as they were at the height of the Scientific Revolution, the solutions to these challenges would be found in the methods of science.[1] Science was discovering nature's secrets in leaps and bounds—in astronomy, biology, and physics—and working miracles in medicine, engineering, and technology. And because science supports its claims with observable, demonstrable evidence, it seemed an especially promising method for resolving entrenched disagreements. The new metaphysical picture of the world that accompanied the new science fit uneasily with the old ways of understanding morality, pushing out these old views and creating space for new approaches.

Thus, the new science and its accompanying metaphysics began to take the place of the medieval picture and methods. But there is much to be gained from a closer look at the details and texture of this transformation.

THE BACKDROP: ARISTOTELIAN SCHOLASTICISM

For the last few centuries of the Middle Ages and into the first of the modern era, a particular school of thought prevailed throughout Western Europe. This was Aristotelian scholasticism, or "scholasticism" for short. Scholasticism was both a broadly shared metaphysical picture of the world and a method of inquiry. It arose in the monastic schools developed during the Carolingian Renaissance of the eighth and ninth centuries and was built from a specifically Christian understanding of reality, expressed in terms of metaphysical concepts devised by Aristotle. The leading intellectuals of the age, including Duns Scotus (1266–1308), William of Ockham (1288–1347), and, most significantly, Thomas Aquinas (1225–1274) embraced, developed, refined, defended, and taught versions of this view of reality, helping establish it as the dominant paradigm of moral and intellectual inquiry.

As a rule, the scholastics sought to understand reality for the purpose of contemplation. Their goal was less to make discoveries and more to understand the things of this world with which they were already familiar.[2] Any given phenomenon was thought to be explainable by identifying four kinds of causes (or explanations), originally postulated by Aristotle, that together make up a complete explanation. These causes were the material, the efficient, the formal, and the final. Consider, for example, a human being. The material cause of a human being is the matter—in today's terms, the carbon, oxygen, hydrogen, and other elements—of which the human is made. The formal cause is the form or essence of humanity that, when combined with the matter, causes that carbon, oxygen, and so on to be something more than just a collection of chemicals: namely, a human being. The efficient cause is that which brought this particular human into existence—for instance, the procreative union of the person's father and mother, and the subsequent development in utero. And the final cause of the human is the purpose or function of humanity: for the scholastics this would have been, roughly, to know, contemplate, and commune with God.[3]

Understanding forms and final causes wasn't thought to be epistemologically problematic. Consideration of a familiar object was usually sufficient to allow someone to understand that thing's purpose or end. That an object *seemed* intended to achieve some end was taken as evidence that this was so.[4]

In the scholastic picture of the world, then, common sense was basically correct. The natural world was constituted by different kinds of objects and qualities that were not reducible to other phenomena. Wind, fire, rain, the earth, a man, a horse, a flower, and so on were qualitatively different realities. They needed to be understood systematically, but a sound theoretical understanding was to build on common sense rather than undermine it.

Following from this, it was the teleological properties of things—their final causes—that provided a framework for an ethics. If something was designed in a particular way for a particular good purpose, then one had some obligation to help realize that thing's true being.[5] Roughly put, moral laws flowed from the nature of things.[6]

The heritage passed down from scholasticism, then, located the source of morality in God, and his moral laws allegedly were revealed in the essences and purposes of things. There were disagreements often enough, but nothing that led to a discrediting of the scholastic methods of inquiry.[7] The absence of widespread, fundamental disagreement about the moral law was no doubt in large part due to the power and reach of Catholic hegemony. Rome had always been able to resolve or quash disputes before they undermined the overarching system; for many centuries, Rome maintained sufficient social, political, and intellectual authority to mitigate if not prevent division. This was not to last.

CONFLICT AND COMPLEXITY

The transition in Europe from the late medieval to the early modern period was a time of significant social, political, economic, cultural, and intellectual changes. These included a growing sense that scholasticism was no longer adequate to make sense of the world.

One of the more significant causes of this sense of inadequacy was the pressure of religious and political conflict. Through the late medieval period, Europeans generally agreed with their neighbors on basic religious principles—agreement that, again, was reinforced by the power of the Roman Catholic Church and the political apparatus of its various client states. There were signs of fissure, however, not least brought on by its own corruption, decadence, and abuse of power. The ecclesiastical reforms of John Wycliffe, Jan Hus, and Girolamo Savonarola in the fourteenth and fifteenth centuries, though unsuccessful, seeded a dissent that finally gained traction at the beginning of the sixteenth century, when an obscure Augustinian monk and scholar named Luther initiated a movement that eventually fractured the authority of the Catholic Church across Western Europe. By the middle of the sixteenth century, Europe's dozens of microstates had adopted a wide range of religious allegiances. A Catholic nobleman recognizing the authority of the pope might have nearby political rivals who were Lutheran or Calvinist, with whom he would disagree over the relative authority of the church and scripture, the holiness of ordinary life, who is qualified to interpret scripture, and other issues.

The source of the conflict was of course mixed, with motivations that were as much political as theological. But the theological disagreements were not insignificant.[8] Differences over what God had revealed to humankind about ultimate truth proliferated and deepened, and in turn, conflict between rival political-religious states escalated into numerous European wars from the early sixteenth century through the middle of the seventeenth century. The Thirty Years War and the French Wars of Religion are well-known events, but sometimes forgotten is the bloodiness of those conflicts. Killing in God's name produced huge death tolls. Upward of 4 million people, out of a total regional population of 16 million, were killed during the thirty-seven years of the French Wars of Religion; 7 million died because of the Thirty Years War, out of a population of 20 million. In all, the wars of religion claimed between one-third and a quarter of those then living in Western Europe.[9] Few families were spared casualties. The urgency to ameliorate such gratuitous destruction was palpable. It simply could not go on.[10]

Another source of pressure was early modern capitalism and the increased trade that resulted from exploration and colonial conquest. What moral principles would underwrite international trade laws capable of governing commercial interactions with peoples from distant lands?[11] This is a different version of the problem of disagreement. After the Renaissance, state budgets swelled, regional commerce grew, international trade flourished, and international finance became more complicated.[12] These developments generated a need for just and ethical contracts to keep trade dependable and civil and tax law universally fair. Exploration only intensified such problems. When the Turks took Constantinople in 1453, the Silk Road that had given Europe access to the wares of India and the East was blocked, compelling European nations to find new trading routes by sea. Richard Tuck put the problem vividly:

The Indian Ocean and the China Sea were an arena in which actors had to deal with one another without the overarching frameworks of common laws, customs, or religions; it was a proving ground for modern politics in general, as the states of Western Europe themselves came to terms with religious and cultural diversity. The principles that were to govern dealings of this kind had to be appropriately stripped down: there was no point in asserting to a king in Sumatra that Aristotelian moral philosophy was universally true.[13]

European states needed an ethics that could bridge both internal, religiously motivated disagreements and external differences with distant trading partners.

Yet another source of pressure was the increasing awareness of the complexity of the natural world. The exposure to peoples, cultures, flora, and fauna previously unknown to Europeans was another consequence of global exploration. The old biological taxonomies passed down for centuries from Aristotle, Pliny the Elder, and Galen were simply inadequate to account for the influx of new specimens and life forms. At the same time, new cultural encounters revealed that many moral and social practices thought universal were in fact merely European customs. Europe was confronted with strange new creatures and tales of distant lands, and Europeans became curious about what else awaited discovery.[14] Clearly, the ancients had not taken the full measure of the world's contents.

A Turning Tide

To many early modern thinkers, the pressures on scholasticism appeared ubiquitous and unyielding. The ethical theories of the scholastics grounded morality in natural laws, which were alleged to be graspable by consideration of the essences of things—that is, by appreciating the ends to which things were made. But these essences weren't so obvious as to compel agreement on the moral laws that followed from them. Religious conflict in Europe demanded ethical systems capable of overcoming disagreement, and contemplative, academic scholasticism began to seem impractical as a foundation for a common legal culture capable of adjudicating complex and competing economic interests.[15] In short, to those concerned with the dilemmas posed by the early modern era, scholasticism seemed an inadequate resource for potential solutions.

The Promise of the New Science

The aims and methods of scholasticism also fit poorly with the aims and methods of early modern science. Here, capitalism was again a factor, as the transition out of feudal systems strengthened the connection between innovation and the exploitation of material resources.[16] Scholastic methods were oriented to fundamentally different goals. As the seventeenth-century English scientist Francis Bacon put it, the scholastics had spun out "cobwebs of learning admirable for the fineness of thread and work, but of no substance or profit."[17] What was needed from scholars was a new commitment "to use and not to ostentation."[18]

The scholastics' logic of inquiry had depended on the Aristotelian syllogism, a method that proceeded by making universal claims, often supported by quotations from prominent figures from the classical Christian tradition, and then using these universal claims as premises from which to draw particular conclusions. A well-known example illustrating the structure of these syllogisms is as follows:

1. All men are mortal.

2. Socrates is a man.

Therefore,

3. Socrates is mortal.

Bacon argued that the scholastics had the relationship between the universal and the particular exactly backward. Starting with allegedly true universal claims fails to allow for the possibility of disconfirming counterexamples. Bacon argued that we should instead begin with extensive observation, taking careful note of the entire range of possibilities. Only after this sort of data had been gathered and studied could one hope to justify a universal claim.[19]

Bacon's inductive method, rooted in careful observation, was only the starting point for the formation of a new science. Active experimentation soon followed. The genius of experimentation was that it could focus on specific hypotheses and provide repeatable tests that anyone could observe to confirm or disconfirm their truth. Its range of application seemed limitless.[20]

The invention and acceptance of new tools, such as the microscope and telescope, and new techniques of examination, such as dissection, generated an abundance of new and exciting insights into things that previously had been seen as commonplace. Among the most important of these were discoveries concerning the workings of the human body. Biologists such as Andreas Vesalius (in *De Humani Corporis Fabrica*) and Amato Lusitano (in *Curationum Medicinalium Centuriæ Septem*) delved deeply into the details of human anatomy, and much of what they found had obvious application in treating human illness.

The Scientific Revolution drew inspiration in part from a renewed interest in the thought and mores of ancient Greece and Rome. This neoclassicism grew first in Italy during the

Renaissance but would then sweep through Europe. One consequence of this renewed interest in classical thought was an increased focus on mathematics and its role in explaining nature.[21] Ancient Platonists had held that mathematics was what was most fundamental or real about the world. Galileo in Italy, Kepler in Bavaria, Boyle and Newton in England, and Gassendi in France, among many others, adopted a similar view, which led them to astonishing innovations in astronomy, physics, and chemistry—innovations rooted in precise mathematical formulation of lawful relationships between basic physical properties.[22] These successes encouraged the application of mathematics to natural phenomena, where it had never been utilized before. As Galileo put it,

> Philosophy is written in that great book which ever lies before our eyes—I mean the universe—but we cannot understand it if we do not first learn the language and grasp the symbols, in which it is written. This book is written in the mathematical language, and the symbols are triangles, circles, and other geometrical figures, without whose help it is impossible to comprehend a single word of it.[23]

The discoveries made possible by representing basic physical qualities in numerical units led to a significant shift in the conceptual resources for describing reality.

The notion that the universe was structured mathematically was akin to the idea that the universe was parsimonious or theoretically elegant. The idea of parsimony, now commonplace in science, is that theories positing fewer concepts or moving parts to explain some phenomenon are more likely to be true than those positing more. Kepler, for instance, assuming that reality was mathematically elegant in this way, sought out and found a simpler model for celestial mechanics.

It is difficult to describe how radical a shift in perspective this was. Against the scholastic view, which held that material entities possessed animating, purposive qualities that explained their movement and behavior, René Descartes argued that the material world was inert, inanimate, lacking mental or experiential qualities, and devoid of inherent purpose. Observable reality was constituted by elements whose features were fully physical and describable in quantitative terms,[24] a claim powerfully illustrated by Gassendi's argument that all features of things could be accounted for in terms of the features of their smallest parts—the *corpuscles*—that composed them.[25] Philosophers in a Cartesian spirit distinguished the so-called primary qualities of things, including extension, solidity, motion, and the like, from their secondary qualities, which included the properties of everyday experience such as the sensory qualities of color, taste, and smell. The "real" world in which human beings live was no longer seen as a world of substances whose ultimate qualities could be directly experienced. It had become a world of atoms and particles equipped only with mathematical characteristics and moving according to laws fully expressible in mathematical form.[26] As Descartes put it, "Give me extension and motion, and I will construct the universe."[27]

No one was more important to the new science than Isaac Newton. In astonishingly simple equations, his *Principia Mathematica* captured the rules that explained the behavior of much of the physical universe. The significance of his discoveries surrounding the movement and interaction of physical objects reinforced the idea that reality was wholly constituted by matter in motion and governed by laws within nature itself. The horizon of scientific possibility stemming from his discoveries opened up, and with it hope for its application.

It should be noted that Newton's work reinforced the increasingly popular mechanistic view of the world, but also tempered it. He relied on mechanical explanations where he could, but he refused to posit mechanisms to explain the new laws his equations described. The equations were true: they were elegant descriptions of the behavior of natural phenomena, and they had astonishing predictive power. But Newton, more fully than those before him, broke with the lingering legacy of the scholastics in that he resisted the impulse to propose speculative explanations for why his equations were true. To see the point more clearly, consider Newton's third law of motion, paraphrased: for every forceful action of one object on another, the second body exerts an equal force back on the first object. A scholastic might have hazarded that this is true because every object possesses *a power* that seeks to respond in a like manner to whatever contacts it. A mechanist might have proposed that all objects have a specific mechanical structure that explains why objects would interact the way the third law describes. But Newton saw such explanations as mere guesses. "Hypotheses non fingo" (I don't feign hypotheses) was his maxim. All that was needed was to represent the world's features quantitatively and then systematize and simplify the mathematical relationships among those features. This basic theoretical logic would be tirelessly emulated by thinkers from Newton's day to our own.

In sum, optimism about the new science was grounded in its success in discovering facts about the world, its usefulness to a large range of problems, its intelligibility compared to often obscure scholastic metaphysics, and its potential for resolving longstanding disagreements through physical experimentation and proof. But the new science also generated optimism because it offered evidence—often in the form of widely observable experiments—that some of scholasticism's most important statements about the natural world were simply false. Galileo, for example, disproved many of Aristotle's well-known claims, showing that the sun has spots, that the moon's surface isn't perfectly smooth, that all massive bodies fall with the same acceleration, and that Venus has phases like the moon.[28]

From the vantage point of the Enlightenment a century later, an opaque philosophical tradition bloated by abstruse speculation *had* to give way to an efficient and observably correct new picture of the world.[29] How could it not? As Holbach scornfully said of the metaphysician in his *Système de la nature* in 1770, "He despises realities to meditate on chimeras; neglects experience to indulge in systems and conjectures; dares not cultivate his reason."[30]

Yet for all we can observe in history—in the views defended and discarded—scholasticism may in principle have had the resources to adopt and adapt to the new science. It is far from clear that the arguments presented against scholastic metaphysics were decisive from a philosophical perspective—or even that compelling.[31] Neither was there a clean break from theologically speculative metaphysics to secular reason. Newton, whose scientific approach has been endlessly emulated, in fact wrote little on science compared to his 4 million words on theological matters.[32] Still, the triumph of the new scientific method and new physics over scholasticism has often been seen as sudden, clean, and inevitable.

The Modern Natural Lawyers of the Seventeenth Century

The expansive theology, ethics, and social structure of medieval Christianity had provided a larger plausibility structure for scholasticism for many centuries. Indeed, Christianity and scholasticism had become so intertwined that their pictures of the world were seen as more or less the same thing.[33] But by the seventeenth century, the credibility of both scholasticism and Christianity among European intellectuals was waning. To be sure, scholasticism didn't disappear overnight: it remained solidly ensconced in European universities for many years following the Scientific Revolution. But the tide had turned.

As the authority of scholasticism began to wane, so did the credibility of ethical accounts grounded in Christian theology. The idea that nature was a purposeful realm ordered by intrinsic teleological dispositions could no longer be sustained. How, after all, does one demonstrate final ends? What are the units of measurement? How does one represent teleological properties mathematically? Since the old ways of thinking about morality (or reality) no longer provided answers, the new scientific methods seemed like a plausible source for an alternative account of morality.

In the early seventeenth century, a number of important philosophers and legal theorists initiated the first self-consciously scientific approaches to morality. The main figures here were Hugo Grotius (1583–1645) and Samuel von Pufendorf (1632–1694).[34] Against the view that moral laws governing humankind were derivable from the evident natures and final ends of the created order, the early natural lawyers began with a view of nature as a nonpurposive realm of atoms on which God imposes, by an act of will, motion and an extrinsic order of efficient causes or regularities.[35] Their task was to bring natural law into line with this concept of nature and human nature, and to do so in ways that located moral laws in those rules necessary for the existence and survival of civil society, rather than in realization of transcendent higher goods.

Grotius and Pufendorf explicitly recognized a need for a moral theory, rooted in scientific objectivity, that could create a stable political society in the face of disagreement, skepticism, and the pragmatic failure of scholastic philosophy.[36] They hoped that by insisting on observable evidence to support moral claims, they would offer a way to temper some of the most violent conflagrations of human unsociability.[37]

Grotius's solution to seemingly ubiquitous disagreement—both within Europe and between European nations and their trading partners in the East—was to identify bases for law and morality that were acceptable to all parties by virtue of their being more or less observably evident[38]—ethical claims so obvious that none could deny them.[39] His argument was rooted in two claims about humanity: that humans are prone to controversy and conflict, and that humans have a desire to live together in society. His moral ideal, then, was to find a way to limit conflict while permitting human sociability. The moral laws were whatever principles could best resolve these conflicting aims. And since we can determine that some resolutions are better than others, the moral laws are empirically discoverable.[40] Because, by Grotius's lights, we will be able to determine which proposed systems of laws best balance our desire for society with our susceptibility to conflict; the laws depend on no other source. He famously said that these laws would hold even if there were no God. This marks one of the first instances of a modern thinker asserting that ethics could be grounded in something other than God.[41]

Grotius saw the epistemological basis for his two claims about humanity as broadly empirical—at least on similar footing as empirical study as it was then understood. "My first care," he wrote in *The Rights of War and Peace*,

> was to refer the Proofs of those Things that belong to the Law of Nature to some such certain Notions, as none can deny, without doing Violence to his Judgment. For the Principles of that Law, if you rightly consider, are manifest and self-evident, almost after the same Manner as those Things are that we perceive with our outward Senses.[42]

Grotius's moral theory thus came from a self-conscious effort to follow some of the major methodological elements of the current science, including a commitment to detailed observation, elegance in formulation of lawlike phenomena, and publicly verifiable bases for key claims.[43] What was Grotius's proposal?

> Laws . . . tell us what is or is not in accordance with the kind of society of rational beings that we all want. They do so in the first instance in a negative way. That is lawful which is not unjust; and to be unjust is simply to violate rights. (I.I.iii.1, p. 34)[44]

The key concept here was "rights," where for Grotius a right was "a moral Quality annexed to the Person, enabling him to have, or do, something justly."[45]

Implicit in Grotius's view was a realist metaethics—which is to say, he thought that what made moral claims true was the existence of real moral properties that actually inhere within every individual.[46] The problem, of course, was that it was never clear how rights could be understood in terms of quantitatively representable physical features, or in terms of any sort of fundamental matter. In the mechanistic world-picture that was rapidly becoming scientifically and philosophically dominant, Grotius's theory had the same sorts of shortcomings as did the old Aristotelian metaphysics.

Transitional Figures

This is where the thinking of Thomas Hobbes (1588–1679) marks a departure. He too was deeply concerned by the conflict of the age. Without coming together to form a peaceable society, he believed, humankind would live in a state of nature—a war of all against all—with everyone individually vying for resources against all others. The bloody wars of religion taking place on the continent, as well as the English civil wars of religion, were grotesque manifestations of such conflict, and doubtless influenced Hobbes's perspective: indeed his great work *Leviathan* was published in the final year of the wars. In such a state, everyone is at constant risk of death. Hobbes's motivation to propose a system of morality that could avoid the perils of disagreement is clear. He had a minimalist view of ethics, viewing mere survival as the chief good for human beings.[47]

Like the other modern natural lawyers, Hobbes hoped to establish an ethics that could win the assent both of rival religious factions and of moral skeptics within a framework that would be describable in terms acceptable to the new science.[48] But if Grotius's picture of objective morality had fit uneasily with the increasingly compelling mechanistic world-picture, what sort of moral theory would do? Here Hobbes could be seen as offering a solution.

He began by denying that morality is an inherent part of the natural world or human nature, or that there is any morality prior to humankind's crafting of laws.[49] As he put it, "For these words of Good, Evill, and Contemptible, are ever used in relation to the person that useth them: there being nothing simply and absolutely so; not any common Rule of Good and Evill, to be taken from the nature of the objects themselves."[50] Instead, the moral law is whatever human beings make it to be through consent and convention.

How then to account for the realm of the moral? Hobbes attempted to reduce all moral motivation—action toward the good and away from the bad—to the movement and interaction of material particles. His basic position was that human action toward something perceived as good can be understood as the movement of all those particles that make up a human body. The cause or impetus for each such movement can be traced back to the material interaction of particles, rather than to an immaterial will.[51] To get a sense of what Hobbes had in mind, think of an individual as being composed of tiny billiard balls. Sensation of an object consists in other billiard balls entering the body through the senses, colliding with some of those inside the body, and initiating a causal sequence of collisions. These collisions spread through the body in various ways, and if enough balls are bounced in the same direction, this constitutes the body moving in that direction. If the body moves toward the object of the sensation that started this series of events, then Hobbes counted this as desire for that object; if away, he counted it as aversion.

The next step was to provide a moral psychology that would

connect human desire and aversion to morality. In scholastic thought, the prevailing idea had been that human beings judged something to be good or bad, and that this explained why they were attracted or repulsed by that thing. Hobbes reversed the order of explanation, arguing that human beings were motivated by their personal interests, and that out of those interests came their understanding of good and evil.[52] The good, for Hobbes, was "whatsoever is the object of any mans Appetite or Desire"; evil was the opposite: "the object of his Hate, and Aversion."[53] In this way, he provided a mechanistic account of good, evil, and human action, introducing one of the earliest modern instances of a theory that would later become one of the most popular scientific approaches to moral psychology.

Hobbes's proposal for how humanity might find peace and avoid violent disagreement—his political philosophy—was partly built on this picture of human morality. As we've seen, Hobbes held that there is no morality except that which we make by contract and agreement, and that good and evil consist only of those things we desire and hate. Add to this Hobbes's belief that human beings naturally—in the absence of strongly enforced laws—are prone to behaviors sabotaging each other's chances for getting what we want, and the best political system should be one that we create by consent to give the state great authority to enforce peace. This is our best chance to pursue what is good and avoid what is bad.

The contribution of John Locke (1632–1704) at this point was critical. His contribution was less toward an explicit scientific theory of morality than it was toward a comprehensive *empiricist epistemology*—an account of human knowledge rooted in sensory experience. Locke put the case this way:

Let us then suppose the mind to be, as we say, white paper, void of all characters, without any ideas:—How comes it to be furnished? Whence comes it by that vast store which the busy and boundless fancy of man has painted on it with an almost endless variety? Whence has it all the materials of reason and knowledge? To this I answer, in one word, from EXPERIENCE. In that all our knowledge is founded; and from that it ultimately derives itself. Our observation employed either, about external sensible objects or about the internal operations of our minds perceived and reflected upon by ourselves, is that which supplies our understandings with all the materials of thinking. These two are the fountains of all knowledge, from whence all the ideas we have, or can naturally have, to spring.[54]

If empiricism was the most effective and accurate means of generating knowledge, then moral knowledge must be pursued through empirical methods. In short, the credibility of empiricism made scientific accounts of morality more plausible.

Locke never did propose a scientific account of morality, though he seems to have intended to and did, in fact, gesture toward one.[55] Like the other natural lawyers of his time, Locke's approach to morality was motivated by the desire to avoid the confessional disputes that marked the age, to persuade moral skeptics, and to fit a mechanistic view of the universe.[56]

A key feature of Locke's groundwork for a moral theory was his novel account of how moral ideas develop. They are not the direct product of experience but are cobbled together from ideas directly gained from experience. Locke called such ideas "mixed modes." Since mixed modes are constructed by human thinkers, they do not represent an independent reality. Instead, they are "archetypes made by the mind, to rank and denominate Things by."[57] And because moral ideas are constructed by human thinkers, we can have precise knowledge of them and can clearly see what can be deduced from them. This is why Locke claimed that moral knowledge was capable of demonstration.[58] As with Hobbes, Locke's account of the moral qualities *good* and *evil* was hedonistic. In other words, he defined "good" as that which brought us pleasure and "evil" as that which brought us pain.[59] On this basis, Locke argued that the authority of morality and law came from what sorts of behaviors would bring us rewards versus which behaviors would bring us punishment. Rewards bring pleasure and therefore are good, while punishments bring pain and therefore are bad. As the Marquis de Condorcet summarized it,

> In the same manner, by analyzing the faculty of experiencing pain and pleasure, men arrived at the origin of their notions of morality, and the foundation of those general principles which form the necessary and immutable laws of justice; and consequently discovered the proper motives of conforming their conduct to those laws, which, being deduced from the nature of our feeling, may not improperly be called our moral constitution.[60]

A Coming of Age

By the eighteenth century, Europe's intellectuals were increasingly confident that humanity had matured to the point of being able to make judgments without recourse to the external authority of tradition, the church, or God. The rhetoric, of course, was always more inflated than reality could justify. Propagandists were more common than intellectuals. But the Enlightenment was clearly gaining ideological momentum, and its accomplishments in math and science were undeniable. Again and again, the new science demonstrated that its method for systematically understanding the natural world was superior to any alternatives. Philosophy also had established itself as a powerful force independent of theology, one capable of challenging the authority and legitimacy of the old regime. This was a *siècle des lumières* in which light was expected to increasingly dispel the shadows of superstition. Enlightenment intellectuals were confident that the world and human experience were fully intelligible to natural reason and that reason alone would lead to a more humane civilization.

The philosophers in Great Britain, the *philosophes* in France, and the *Aufklärer* (enlighteners) in Germany were those who, in the words of Diderot's *Encyclopedia*, "trampl[ed] on prejudice, tradition, universal consent, authority, in a word, all that enslaves most minds, dares to think for himself, to go back and search for the clearest general principles, to admit nothing except on the testimony of his experience and his reason."[61] The tone and ambition of these intellectuals and their audiences varied from Paris to Berlin and from Edinburgh to Philadelphia, but they shared a sense that the "dread and darkness of the mind . . . require not the rays of the sun, the bright darts of day; only knowledge of nature's form dispels them."[62] The promise of emancipation from the bondage of ignorance, superstition, and error seemed ready to be fulfilled.[63]

An important shift had taken place in their understanding of the ends or purposes of morality. For ages, it was taken that the highest end of humanity was to contemplate God and to perfect the natures of individuals, both alone and in society, in accordance with divine intention. Moral law comprised the principles that tended toward the fulfillment of these ends. But these teleological ideals had been abandoned by the new philosophy. Now the end of morality was happiness in this life, rather than union with God or perfection of a God-given telos. A worldly, secular ethics grounded in a broadly naturalistic foundation was replacing an otherworldly, theologically based ethics.

Toward this end, what religion and metaphysics had made obscure, now science must illuminate. In this spirit, the *philosophes* continued to seek new sources for moral theory, now within the ambition of a comprehensive science of man and nature.[64] In this larger endeavor, the study of morality was to proceed in much the same way as any other scientific study.

Underwriting this optimism was a philosophical naturalism, a sense that the only things that enter into explanation are empirically describable, physical things.[65] Here, as in many other arenas of thought, Locke's empiricism was taken as a guide. In an account of the growth and progress of rational thought in humankind published in 1794, Condorcet observed that Locke had "grasped the thread by which philosophy should be guided . . . that all ideas are the result of the operations of our minds upon sensations we have received." For this reason, he wrote, Locke's empiricism would soon be "adopted by all philosophers" with the hope that by applying it *"to moral science, to politics and to social economy,* they [would be] able to make almost as sure progress in these sciences as they had in the natural sciences."[66]

At least initially, Locke's ideas made the science of humanity an extension of physical science.[67] D'Alembert wrote that Locke had "reduced metaphysics to what it ought to be in fact, the experimental physics of the soul."[68]

If Locke's empiricism was the proper guide, Newton's achievement was the ideal. Following Newton, the goal of the science of man was to identify the basic features of human moral life and the laws or principles that related them, so as to permit a general explanation of moral phenomena. This research program had greater ambitions than mere understanding. Once moral scientists had achieved a Newtonian level of explanation, the predictive power and control that were becoming possible in the physical sciences would become available in the moral sciences as well. This would make possible a universal practical science that would transform law and politics.[69]

Three Schools of Enlightenment Thinking
AND ONE LINGERING AND DEEPLY
DISTURBING WORRY

AND SO IT WAS that the Enlightenment spawned optimism about the power of human reason to systematically comprehend the natural world and serve as an authoritative guide to the practical realities of human affairs. By the eighteenth century, this optimism extended fully and explicitly to the ideal of building a moral science of humanity and, in this, a scientific foundation for morality itself.

Over the next two centuries, important differences would emerge in the strategies for building a scientific basis for morality. The differences concerned the specific way each strategy attempted to connect the moral realm with the empirical realm. The three most influential strategies were sentimentalism, utilitarianism, and evolutionary ethics.

- *Sentimentalism* identifies basic moral phenomena— good, bad, right, wrong, etc.—with feelings or sentiments, rather than with things or actions themselves.
- *Utilitarianism* begins with Hobbes's and Locke's understanding of good and bad in terms of pleasure, and adds

that all ethics consists of is calculating how much pleasure or pain actions produce.

▸ *Evolutionary ethics* recognizes that human beings were produced by evolution and tries to connect moral progress to evolutionary development.

Over time, these strategies for thinking about science and morality have been developed and refined into different traditions that define the parameters of constructive thinking about how a scientific foundation for morality could be built. For instance, while the fact of human evolution remains of paramount importance for the science of morality, no one now attempts to equate moral development with evolutionary development. But even in its innovations, the science of morality has tended to operate within the rough boundaries set by these strategies.

SENTIMENTALISM

Sentimentalism emerged in the eighteenth century partly in response to the sense that some of the leading moral theories of the day did not adequately capture human nature in explaining moral motivation—why human beings have an inclination to act rightly. In one view—the egoist view[1]—human beings were primarily motivated by self-interest; in another view—the rational intuitionist[2]—human beings understood their obligations through the rational discernment of moral properties. But were human beings merely self-interested? And were they oriented toward moral life primarily through their rational faculties? The sentimentalists doubted both theses. They thought a true science of man would reveal that human beings were often moved to act rightly out of benevolence as well as egoism—that unselfish affection played an important role in explaining the moral life.[3] They also believed that the passions were at the heart of the moral life.

Several important figures laid the groundwork for the sentimentalist account: Anthony Ashley Cooper, better known as the Third Earl of Shaftesbury (1671-1713); Francis Hutcheson (1694–1746); and, of course, Adam Smith (1723–1790) among them.[4] But arguably the central figure was David Hume (1711–1776).

Living in the heart of the eighteenth century, Hume understood the pressure generated by the reigning empiricism of the day. He fully embraced that ideal and sought to understand moral thought and feeling through observable phenomena. The subtitle of his *A Treatise of Human Nature* was *Being an Attempt to Introduce the Experimental Method of Reasoning into Moral Subjects.*" In the volume's Introduction, he wrote:

> Tho' we must endeavour to render all our principles as universal as possible, by tracing up our experiments to the utmost, and explaining all effects from the simplest and fewest causes, 'tis still certain we cannot go beyond experience; and any hypothesis, that pretends to discover the ultimate original qualities of human nature, ought at first to be rejected as presumptuous and chimerical.[5]

Yet Hume also observed that unlike more exacting sciences, the science of morality could not rely on more rigorous experimentation. Instead, "We must therefore glean up our experiments in this science from a cautious observation of human life."[6]

Hume wanted to do for the mind what Newton had done for mechanics. As Andrew Janiak puts it, "For Hume, this meant following what he took to be Newton's empirical method by providing the proper description of the relevant natural phenomena and then finding the most general principles that account for them."[7]

Perhaps the two most prominent features of Hume's philosophical work are his naturalism and his skepticism.[8] His naturalism can be seen in his radical attempt to rely only on empirical knowledge, rather than on appeals to reason or metaphysical intuition. His skepticism follows. How? In Hume's view, all of our knowledge comes from perception. "Nothing is ever present to the mind but its perceptions; and that all the actions of seeing, hearing, judging, loving, hating, and thinking, fall under this denomination."[9] For Hume, we don't have any immediate experiential contact with reality—all we have are our perceptions of reality. Perception, then, mediates reality for us.[10] True empiricism, for Hume, meant that we can know only our perceptions, not the actual objects the perceptions purport to be about.[11]

The skeptical turn in Hume's empiricism had important consequences for the evolving understanding of morality.[12] Consider the contrast between Shaftesbury and Hume. For Shaftesbury, there was an objective—that is, mind-independent—order of value that could be comprehended through reason. In his view, that logic was teleological. The goodness of an action or event or phenomenon was that which contributed to the well-being of the whole of which it was a part. In this light, moral sensibilities—the human predilection to evaluate and judge things morally—reflected and gave expression to that moral order. Hume's empiricism, by contrast, led him to conclude that moral evaluation simply expressed a person's feelings and attitudes with respect to a person or situation. Such sentiments were the sum and substance of moral life, and they could never be connected to any objective, mind-independent moral order.[13]

So how does moral evaluation actually come about? According to Hume, reason is necessary for understanding the facts of a situation or dilemma and for tracing out the potential consequences of a course of action. Yet reason by itself cannot determine what is virtuous or vicious; in fact, no moral conclusions could ever be inferred by factual premises alone. This is "Hume's Law": no ought from is. Nor, he argued, could reason move us to moral action in the way that sentiment could. As he famously put it, reason is the "slave to the passions," beholden to such sentiments as fear, desire, repugnance, hope, joy, love, hate, and the like. "To have the sense of virtue," he wrote, "is nothing but to feel a satisfaction of a particular kind from the contemplation of a character."[14] Thus,

> when you pronounce any action of character to be vicious, you mean nothing, but that from the constitution of your nature you have a feeling or sentiment of blame from the contemplation of it.[15]

And:

> An action, or sentiment, or character is virtuous or vicious; why? Because its view causes a pleasure or uneasiness of a particular kind. In giving a reason, therefore, for the pleasure or uneasiness, we sufficiently explain the vice or virtue.[16]
>
> [These] moral distinctions depend entirely on certain peculiar sentiments of pain and pleasure.[17]

In short, good and evil are not unlike heat and cold. We don't have knowledge of them except through certain feelings. We come to understand the moral good by experiencing the pleasure of social approval, and we understand evil by experiencing the uneasiness and pain of disapproval. Our experience can be physical or psychological and direct or indirect, as in the emotion we imagine experiencing when we contemplate a person, a situation, or an action. A trait gives rise to approval when it is immediately agreeable to the person who has it or to others, or is beneficial to its possessor or to others. Our understanding of virtue and vice, then, turns on our emotional response to an action, situation, or character. What, then, does morality allow and what does it condemn? This is now an empirical question: when we learn what people morally approve and disapprove of, then we find the claims of morality.[18]

On the face of it, this would seem to lead to a facile relativism, for people clearly differ in what they approve and disapprove of. But Hume downplayed both moral error and moral disagreement. Humanity, he argued, is united by a shared disposition among individuals to agree in their sentiments of approval and disapproval. This disposition or ability to agree he calls "sympathy." Consider, he asked,

> the nature and force of *sympathy*. The minds of all men are similar in their feelings and operations; nor can any one be actuated by any affection, of which all others are not, in some degree, susceptible. As in strings equally wound up, the motion of one communicates itself to the rest; so all the affections readily pass from one person to another, and beget correspondent movements in every creature.[19]

Through both our resemblance to others and our proximity to others, sentiments are sympathetically communicated and shared. This varies, of course; the more we resemble others and the closer we are in proximity, the greater the sympathy. The converse is also true. In the end, however, no one forms moral judgments in isolation but rather in an interaction with others as we sympathize with the effects of certain persons or actions on those around us. We are all mutually influenced by the moral evaluations of others. This is how a common moral understanding comes into being.

Over time and through continual and extended social intercourse, human societies develop complex moral systems. Hume divided the virtues (and thus vices) into two types. "Natural" virtues were not dependent upon a formal society but arose organically and cooperatively within small associations. These included benevolence (by which he understood generosity, humanity, compassion, gratitude, friendship, fidelity, zeal, disinterestedness, and liberality), greatness of mind ("a hearty pride, or self-esteem, if well-concealed and well-founded"), and such natural abilities as wit, prudence, eloquence, and good humor. Beyond these were artificial virtues, which arose in impersonal circumstances that find expression in the conventions that make cooperation toward the common good possible. These included honesty with respect to property (or equity), fidelity to promises, allegiance to one's government, conformity to the laws of nations (for political leaders), modesty and chastity (for women), and good manners. "All these are mere human contrivances for the interest of society."[20]

Though this account of morality provided a naturalistic analysis of its nature and components, Hume's subsequent influence took a circuitous path. But in his own time, his account of justice crucially shaped Adam Smith's new theory of the foundations of jurisprudence[21] and his account of morality in terms of human pleasure and pain inspired Jeremy Bentham (1748–1832) to craft the paradigm version of utilitarianism.

UTILITARIANISM

When Hume observed that "moral distinctions depend entirely on certain peculiar sentiments of pain and pleasure," he gave a rather large role to the pleasure and pain that moral agents feel when considering the character traits of others. These reactive feelings, he thought, revealed what human morality was really oriented around: virtuous and vicious character traits.

Bentham, too, wanted a science of morality, but while he took inspiration from Hume,[22] what he gathered from the text was not what Hume had left there. Bentham said of Hume's *Treatise* that "the foundations of all virtue are laid in utility, is there documented," and that "no sooner had I read that part of the work which touches on this subject, than I felt as if scales had fallen from my eyes."[23] Although humans might habitually judge good and bad in terms of character traits, Bentham thought this practice wasn't defensible. Instead, he reasoned that morality ought to be based on the pleasurable and painful consequences of actions, and little else.

Thus, Bentham sought to do away with Hume's trait-focused approach. But in making what ought to be done solely about what brings about pleasure and prevents pain, he also sought to do away with an even older understanding: that some things were prohibited or compulsory regardless of how much pleasure might result or pain avoided by doing otherwise. For instance, in the old scholastic view, the nature of human beings was thought to reveal *natural moral laws*—that they must be treated in certain ways—and on a Grotian picture, it might be that human rights set limits on what one could permissibly do to another human. While Bentham wasn't the first to propose utilitarianism, the way he did it formed the prototype for future such theories.[24] Bentham's elegant formulation boiled morality down to a single principle: "The principle of utility is meant that principle which approves or disapproves of every action whatsoever according to the tendency it appears to have to augment or diminish the happiness of the party whose interest is in question."[25]

Why did Bentham redescribe morality in these new terms? What explains his formulation of utilitarianism? Part of the answer is that more than many other moral theorists of his day, Bentham was a reformer. He hoped to provide a moral system that would support passage of new and better laws in order to make life better for his fellow citizens.[26] Morality and legislation had a similar basis, he believed, in that both were part of morality in some wide sense. The wide sense of morality was "the art of directing men's actions to the production of the greatest possible quantity of happiness, on the part of those whose interests are in view."[27] The key difference between morality in the narrower sense and legislation was a matter of whose actions were directed. When an individual directed her own actions, this was morality in the narrow sense. But when an individual or group of individuals directed others' actions, this was a matter of legislation.

Bentham saw that certain popular moral theories of his day—especially virtue theories—took the fundamental basis of ethical evaluation to be human character. But if such a view were true, discerning moral standing would be difficult and at times impossible. After all, if character is the target of moral evaluation, and character often is not discernible, then basing any sort of civil code on virtuous character will be pragmatically difficult.[28] Acts are easier to evaluate. Bentham therefore proposed a moral theory that took *action* rather than *character* as the fundamental basis of moral evaluation. From the standpoint of policy and jurisprudence, laws that turn on acts rather than character are easier to enforce. So in the eyes of some Enlightenment thinkers, a virtue of utilitarianism was that unlike any previous moral theory; it permitted assessment of right and wrong in empirical terms.

Bentham's ambitions were not restricted to his desire for practically enforceable laws; his aspiration to scientific methods of inquiry played a large role as well. In one way, Bentham was more scientific than Hume. Hume went beyond a mere science of human morality: his reflections on philosophical first principles in the first two books of the *Treatise* led him to the brink of radical skepticism. But Bentham was resolutely uninterested in such purely philosophical considerations. As Douglas Long put it, they

> merely reflected the seductive nature of introspection and subjectivity for sensitive and gifted philosophical minds. One sometimes senses that Bentham almost *feared* introspection. Clearly he was quite determined, in his philosophical work, to look "outward" from the human subject to the "external" world of actions and events. His arena was the "universal system of human actions," and his search was for an external, demonstrable standard to govern it.[29]

Bentham also rejected any reliance upon intuition or an inner, moral "sense." He called such reliance "ipse-dixitism," from the Latin *ipse dixit*, meaning "he, himself, said it,"[30] and felt that beliefs based upon intuition or moral sense had no evidential support other than the word of the one who invokes them. They were useless foundations for knowledge. Where belief is based upon these sources, he wrote, "Morals is what a gentleman pleases. Every man dreams he understands morality and wishes not to be awakened."[31]

Bentham took from Hume the centrality of utility in morality, but unlike Hume, he said, "I see not . . . what need there was for the exceptions." Claiming to see in human moral behavior nothing more than responsiveness to pleasure and pain, he devoted considerable effort to a comprehensive analysis of the language of metaphysics and believed he had reduced all important terms either to fictions or to underlying physical phenomena.[32] On these grounds, Bentham was convinced that moral properties—goodness, obligation, rights—were reducible to utility; to terms amenable to scientific study.[33] Through this reduction, he believed that it was possible to provide a more precise classification and quantification of the basic moral feelings[34]—hence Bentham's *felicific calculus*: the classification of kinds of pleasures and pains, and the quantification of their values. He thought that the value of a given pleasure or pain would vary with its intensity, duration, certainty, and proximity, and with respect to whether the experience of it was followed by more of the same kind. In this way, moral theory, which had been a subjective and contestable enterprise for centuries, could now admit of objective, empirical standards.[35]

Ultimately, Bentham found the notion of a *principle of association* useful for explaining how feelings of simple pleasures could be united into a sum that constitutes human happiness.[36] Pioneered by Hume and other earlier Enlightenment thinkers, *associationism* was a psychological view of how ideas related and produced new ideas. Just as Newton had explained mechanics purely in terms of physical objects and their physical properties, without appeal to supernatural or agential influence, so too the associationists hoped to explain thought purely in terms of ideas, their properties, and their relations to each other. Bentham needed to be able to claim that feelings of pleasure across many individuals would converge. If one person's feelings of pleasure differ too much from those of another, it becomes hard to plausibly maintain that the right thing for each person to do is that which promotes the most overall happiness. Why should I do what promotes others' happiness if it doesn't promote my own happiness as well? Bentham's solution was to draw on observations made by earlier moral scientists. Hartley had said that "association tends to make us all ultimately similar; so that if one be happy, all must."[37] Again drawing on Hume's pioneering work in moral psychology, Bentham and his followers saw *sympathy* as the mechanism by which humanity could be united in the object of its happiness. Sympathy was taken to be a naturally occurring and universally present feature of humanity whereby each individual can share standards of pleasure and pain.[38] Because of the sympathy that each human bears to the others, it is supposed to become plausible that what makes each one happy aligns with what makes all happy.

Further Developments in Mill and Sidgwick

Around the beginning of the nineteenth century, the early utilitarianisms of Bentham and others came under severe criticism. One major criticism was that utilitarianism offered an impoverished conception of human pleasures and values. Bentham analyzed pleasure into just two basic components: duration and intensity. As a result, he thought we could easily compare pleasures and pains against each other so as to determine which pleasures were the most worth pursuing. Whichever pleasure was most intense and lasted the longest was the best. By reducing value in this way, Bentham came to think that the simple child's game of push-pin was equal in value to music and poetry, so long as the intensity and duration of pleasure brought by both are the same.[39] Philosophers objected that this conception obscured many significant distinctions between pleasures that transcended mere intensity and duration. Thomas Carlyle, for example, called Bentham's utilitarianism "pig-philosophy," since it seemed to place human experience on a par with those of any creature capable of experiencing pleasure.[40] Bentham, it seemed, had failed to provide an adequate definition of morality.

John Stuart Mill (1806–1873) proposed a new version of utilitarianism designed to weather these criticisms. Like others of his time, Mill thought the only way to determine what happiness consisted of was through scientific investigation, so he did not attempt to intuit basic moral principles apart from experience. Instead, he began with what he took to be the most essential tenets of commonsense morality and inductively derived the principle of utility as the law on which they must ultimately rest. Mill put the principle of utility like this:

> The creed which accepts as the foundations of morals "utility" or the "greatest happiness principle" holds that actions are right in proportion as they tend to promote happiness; wrong as they tend to produce the reverse of happiness. By happiness is intended pleasure and the absence of pain; by unhappiness, pain and the privation of pleasure.[41]

His argument or "proof" of this principle proceeded roughly as follows. The only proof of a thing's desirability is that it is desired. Everyone desires happiness—and *only* happiness—for its own sake (anything else that is desired is desired as a means to happiness).

Since everyone desires happiness for its own sake, happiness for its own sake is all and only what is desirable for humanity in general. So, since happiness is all that is worth desiring, it only makes sense to evaluate actions in proportion to how much or little they realize happiness.

Mill claimed that the only way to determine which pleasures were better than others was by figuring out which pleasures were preferred by most people who had tried all of the pleasures in question. These sorts of appeals to actual human preference would show that people tend to prefer the higher pleasures—those that require greater use of distinctively human capacities. The empirical evidence, he wrote, showed that music and poetry were of greater value than push-pin.

Reinforcing the broader utilitarian account of morality in the spirit of scientific inquiry was Henry Sidgwick (1839–1900). In his book *The Methods of Ethics* (1874), Sidgwick set out to "discuss the considerations which should . . . be decisive in determining the adoption of ethical first principles."[42] He sought a "rational procedure by which we determine what individual human beings 'ought'—or what it is 'right' for them—to do."[43] His method of inquiry was intentionally designed to resemble scientific inquiry in certain broad, structural respects. For instance, he took the commonsense moral views of nineteenth-century Britain—on which there was broad agreement—to constitute data for a theory.[44] Similarly, he sought to systematize ethics, much as scientists had systematized certain empirical domains of study.[45]

At the same time, however, Sidgwick introduced doubt about the empirical basis for ethics, ultimately arguing that utilitarianism required *nonempirical,* intuitive justifications for its basic claim.[46] While not strictly scientific, this claim was still purportedly objective and thus better able to provide a more decisive resolution to ethical problems than the Christian popular ethics of earlier generations.

EVOLUTIONARY ETHICS

While the Enlightenment was waning at the end of the eighteenth century, its intellectual ambitions certainly didn't end with it. The explosion of scientific knowledge, the astounding advances in industry and technology, and the growth in the power and reach of the British Empire set against a backdrop of confidence in the trajectory of human progress all made fertile ground for bold new theories of human understanding. To this point, one of the shortcomings of the Enlightenment quest to formulate a scientific foundation for morality was the failure to explain *why* we have the moral thoughts and feelings that we do. But evolutionary theories of species development would permit deeper scientific explanations. This was part of the promise of Charles Darwin's (1809–1882) groundbreaking works, *The Origin of the Species* (1859), *The Descent of Man* (1871), and *The Expressions of the Emotions in Man and Animals* (1872).

Darwin's extensive observations as a naturalist and his systematic account of the evolutionary development of all living things in a theory of natural selection depicted humans as part of the same order as all other life. Humankind, long held to be categorically unique and distinct from the broader animal kingdom, could now be seen as continuous with it. Human attributes and behavior should therefore admit of the same sort of natural explanation as the attributes of any other species. Morality was no exception. Like any other biological capacity or behavior, human morality must be somehow explicable as an adaptation to environmental conditions.[47] This was the promise.

Darwin's own account of morality was influenced by William Paley's work in *Principles of Moral and Political Philosophy* (1785), in which Paley argued, "Whatever is expedient is right. But then it must be expedient on the whole, at the long run, in all its effects collateral and remote, as well as in those which are immediate and direct."[48] Inspired by this idea, Darwin wrote that goodness of action can be understood in terms of those that promote the preservation of the species.[49] These actions include those that promote sociability, nurture and care of the young, and similar behavior.[50]

But where do these actions come from? Darwin posited that human beings shared natural *social* predispositions with other species and that, on the basis of these, an individual would "inherit a tendency to be faithful to his comrades, and obedient to the leader of his tribe."[51] Moreover, he would "from an inherited tendency be willing to defend, in concert with others, his fellow-men; and would be ready to aid them in any way which did not too greatly interfere with his own welfare or his own strong desires."[52]

Among the inherited predispositions was an "instinctive sympathy" that causes humans to be "influenced in the highest degree by the wishes, approbation, and blame" of others "as expressed by their gestures and language."[53] These develop into deeply rooted dispositions to conform to social practices and to live by the rules that govern human relations. According to Darwin, the "more persistent" social instincts eventually "conquer" the "less persistent," selfish ones.[54]

Darwin argued that all of these dispositions confer survival advantages at a group level; those who acted on them helped their group's chances of survival, relative to groups whose members were more selfish. By locating the basis of morality in other-regarding impulses and behavior, he wrote, "The reproach of lay-ing the foundation of the most noble part of our nature in the base principle of selfishness is removed."[55]

Darwin's moral theory was at best inchoate; his account described only the earliest stages of the development of a "moral sense." But it distinguished itself from utilitarianism in important respects. The emphasis on evolved "instinct" stood in contrast to the utilitarian view that the moral sense developed through learning and calculation. And though often associated with the pleasure principle, his theory did not *require* an association of goodness with pleasure.[56] Darwin argued, "The imperious word *ought* seems merely to imply the consciousness of the existence of a persistent instinct, either innate or partly acquired, serving him as guide, though liable to be disobeyed."[57] Against the idea that moral action was an automatic consequence of the social impulse, Darwin contended that the intellect enabled a human being to consider his or her moral instincts and decide whether or not to follow them. Nevertheless, for Darwin, the intellect, just as much as the social instincts, was a product of evolution.[58]

Darwin's paradigm offered a mechanism, natural selection, that simultaneously made the evolutionary account of biological development broadly intelligible and situated humanity firmly within the biological order. It suggested how human morality might have its basis in evolved characteristics. Darwin, however, never developed a full-fledged ethical theory based on natural selection.

The theory he laid out, however, was enormously suggestive, and it provided a foundation upon which many would attempt to build a naturalistic theory of morality. Among those who put forth theories identifying the human good with evolutionary progress were Leslie Stephen, W. K. Clifford, perhaps most famously Herbert Spencer (1820–1903), and later, Edward Westermarck (1862–1939).

Stephen, noting that any society's survival was partly dependent on its observance of rules of conduct and interaction, argued that Darwinian evolutionary theory could provide justification for the accepted and dominant morality. He thought that these rules were the bulk of morality, that "morality is the sum of the preservative instincts of a society."[59] But why, he asked, should individuals work together for the common good?[60] He didn't think Darwinian theory offered any contribution to ethics that would result in revision to the accepted morality. Instead, its value lay in explaining the origin, development, and value of humanity's moral sense.[61]

Unlike the moral scientists of the Enlightenment, Herbert Spencer was not content merely to provide a theory of our moral feelings and thoughts or an account of how moral sentiments motivate us. Spencer, deeply embedded in Victorian social mores, also wanted to use evolutionary theory to justify the existing morality—to explain why it is that human moral principles and judgments were true.[62] Evolution, for Spencer, provided an account of humanity's slow development from savagery to its height in Victorian civilization.

Spencer's writings on social evolution owed more to Malthus and Lamarck than to Darwin. Indeed, his earliest work on the subject predated *On the Origin of Species* by seven years. Spencer and Darwin admired each other's work and corresponded throughout their adult lives. Not incidentally, Spencer was also a close friend of Thomas Huxley and John Stuart Mill. In his view, the study of society and its morality could become scientific only when it was based on the idea of natural evolutionary law. Indeed, change in all aspects of the universe was subject to the laws of evolution, and though there were limitations, social evolution was roughly analogous to biological evolution: it moved from simple to complex, from undifferentiated to differentiated, in response to the natural and social environment.

Spencer's ethical theory was intensely individualistic and utilitarian. As a utilitarian, he viewed happiness as the highest moral ideal. Yet as it did for Mill, liberty provided the conditions for optimal happiness. Behavior "restrained within the required limits, calling out no antagonistic passions, favors harmonious cooperation, profits the group, and, by implication, profits the average of individuals." In short, wherever freedom as a precept of justice is extolled and practiced, humans thrive.

As to the "science of right conduct,"

> I conceive it to be the business of moral science to deduce, from the laws of life and the conditions of existence, what kinds of action necessarily tend to produce happiness, and what kinds to produce unhappiness. Having done this, its deductions are to be recognized as laws of conduct; and are to be conformed to irrespective of a direct estimation of happiness or misery.[63]

This perspective made him hostile to the socialism of his day—and to any government intervention in human affairs beyond protection from foreign adversaries. Such interventions interfere with the natural evolutionary process.

Edward Westermarck came a generation after Spencer, but in the same tradition. By the early twentieth century, anthropology and sociology had begun to reject the linear and progressive evolutionary ideas of societal development featured in the cruder versions of social Darwinism. Westermarck was an exception. Influenced by Adam Smith, he spent much of his life doing fieldwork searching for universal human moral sentiments. He wanted to map the human variability in these sentiments, which he did by distinguishing between a culture's explicit moral practices and the general moral principles and emotions that underlay them.[64]

For Westermarck, whether a given action was right or wrong was relative to the culture practicing it. Variation in environmental factors and nonmoral beliefs explained how the same universal human moral emotions could lead to widely diverse behaviors.[65] By showing that a practice most people in Western cultures would find abhorrent (such as patricide) can be rooted in shared moral emotions (respect and care for one's parents), Westermarck pointed toward an appreciation of the diversity of human moral practices that was still linked to common evolutionary origins, and that created room for underlying objective moral truths.[66]

ENDURING PROBLEMS

By the beginning of the twentieth century, interest in evolutionary ethics had begun to decline, for several reasons. First, evolutionary theory was in turmoil because it lacked a mechanism to explain evolutionary development. Darwin's own proposal, natural selection, had not yet won the day, and it wasn't clear that it would do so. With central components of evolutionary theory in question, people understandably paid less attention to its implications for ethics during this time. Second, various academic disciplines began to fragment through specialization, making it less likely that biologists studying evolution would engage in philosophical theorizing about the relation of evolution to ethics.[67] Third, belief in the inevitability of human progress underwent a sea change. Nineteenth-century accounts of evolutionary ethics often had a strong progressive bias. But the staggering human causalities of the first world war, made possible by technological progress, made this sort of optimism about human ethical development much more difficult.[68]

There were philosophical reasons as well.

One was that it wasn't clear how some crucial aspects of morality could have developed through the evolutionary process. For instance, as Alfred Russel Wallace argued in 1870, "It is difficult to conceive that such an intense and mystical feeling of right and wrong (so intense as to overcome all ideas of personal advantage or utility), could have been developed out of accumulated ancestral experiences of utility."[69] St. George Mivart, a prominent English biologist, similarly argued that while it was possible to give evolutionary accounts of how animals came to behave in ways that are pleasurable, valuable, or conducive to the thriving of their species, such behavior, to count as genuinely moral, would additionally have to be done with the intention of doing what is right *because* it is right. Evolutionary explanations, he argued, could offer no way of explaining how the additional element of moral intention came to accompany mere instincts toward pleasurable or fitness-enhancing action.[70] In 1905, Theodore de Laguna claimed that evolutionary ethicists "committed the dangerous error of conceiving the significance of morality as exhausted in its material conditions . . . confusing the external limits of morality with its inner content."[71]

Another objection concerned the autonomy of ethics. Sidgwick spoke for many when he wrote that "the investigation of the historical antecedents of [moral] cognition, and of its relation to other elements of the mind, no more properly belongs to Ethics than the corresponding questions as to the cognition of Space belong to Geometry."[72] T. H. Huxley drew much the same conclusion in 1893, writing that "evolution may teach us how the good and the evil tendencies of man may have come about; but, in itself, it is incompetent to furnish any better reason why what we call good is preferable to what we call evil than we had before."[73]

Critics continued to pile on. Finally, in 1903, G. E. Moore famously argued that evolutionary ethical theories committed what he called the "Naturalistic Fallacy."[74] The Naturalistic Fallacy is taking or positing some naturalistic property—a property studied by the natural sciences—to be the meaning of the term "good." According to Moore, many forms of evolutionary ethics took the meaning of "good" to be some naturalistic property, such as "highly evolved" in Spencer's case. This was a fallacy, Moore claimed, because even if the naturalistic property in question happened to coincide with goodness—if, say, all and only those things that were highly evolved happened also to be those things that were good—nevertheless, that doesn't show that the *meaning* of "goodness" *is* the naturalistic property. The coincidence of the terms may be just that—a coincidence.

Moore supported his diagnosis with what is called the "Open Question Argument." This argument aims to show that any analysis of goodness in terms or ideas other than goodness itself will be open to a certain sort of doubtful question, which then shows that the analysis has failed. To see what Moore had in mind, let's consider again the example from Spencer—the claim that "good" means something like "highly evolved." This analysis is open to the question "But is it good to be highly evolved?" The question is significant. But if the analysis were correct, questioning it in this way wouldn't make sense. Consider, for instance, this correct analysis: a vixen is a female fox. It is pointless to ask, "But is a female fox a vixen?" Such a question seems obtuse, since the statement "a female fox is a vixen" is unquestionably true. Any *correct* analysis of a concept, Moore argued, isn't open to this sort of question. But since any analysis of good in nongood terms—including analyses in naturalistic terms—is open to such questions, any analysis of good in nongood terms must be incorrect.

Moore's objection to evolutionarily grounded ethics seems to have been something of a bombshell, and it was a major intellectual factor discouraging philosophical pursuit of evolutionary accounts of ethics for the next seventy years.

In the decades following Moore, ethicists accepted that ethical properties were not naturalistic, but for slightly different reasons from his. Moore argued that any supposition that ethical properties were naturalistic would fail the open-question test. Later, ethicists instead thought ethical properties must be analyzed in terms of goodness, and that goodness—if it existed—would intrinsically exert action-guiding force. But no natural entities—atoms, charges, mass, and so on—can exert this kind of force. So ethical properties must not be naturalistic.[75]

These arguments raised formidable objections to the quest to find a scientific foundation to morality. Against them, the idea that ethics or morality was grounded in features of the natural world and could be studied scientifically lost plausibility.

Meanwhile, the debates over how to formulate and understand utilitarianism had, by the middle of the twentieth century, sapped hope that utilitarianism would yield any sort of objective resolution of ethical issues.[76] While this conclusion did not undermine utilitarian theories in a general sense, it did make evident that there was no objective, consensus-establishing method for figuring out which version of utilitarianism was correct.

An Important Shift in the Methods of Inquiry

The most significant change in the quest to find a scientific foundation for morality at the start of the twentieth century had nothing to do with the concepts of scientific inquiry, but rather with methods.

One of the distinguishing characteristics of intellectual life before the late nineteenth century was ambiguity in the practices of scholarly inquiry. One was not merely a philosopher or a biologist or a chemist or an anthropologist. The pursuit of knowledge was not so neatly divided into self-contained disciplines. This changed at the end of the nineteenth century with the emergence of specialized and professionalized disciplines of inquiry. Anthropologists, to take one case, began to answer their questions without engaging the methods and theories of philosophy or theology. This turn was partly due to an effort to achieve greater rigor as well as scientific clarity and authority.

The case of psychology illustrates not only the general transformation toward a more scientific approach but also the changes taking place in the science of morality. In his *Principles of Psychology* of 1890, William James articulated a new experimental approach to the study of the human mind. His treatment included discussion of consciousness, emotion, the will, and other empirically intangible entities, but because these entities were difficult to describe or study via empirical experiment, many of those pursuing a science of the mind a generation later would exclude them from theoretical treatment. Traditional philosophical approaches to mental phenomena, which often took these intangible mental entities seriously, were thus seen as less and less relevant to psychological research. If psychology was to be a genuine science, then it must proceed by empirically tractable experiment and resist the a priori or introspective speculations of philosophy.[77]

At roughly the same time, psychologists began conducting their inquiry by studying animal behavior. The recently established idea that the human species was continuous with the animal kingdom came with the corollary that animal behavior could offer insights into human behavior. This eased the transition toward an exclusively empirical approach to psychology, since mental phenomena were unobtainable in animal studies, and these studies produced important truths about animal behavior—perhaps most notably, the groundbreaking experiments by Pavlov and others demonstrating the role of conditioning.[78]

John Watson took the empiricist sentiment in psychology to its limit. In a polarizing speech at Columbia University in 1913, which founded behaviorism as a unified research program, he argued that if psychology would be a legitimate science in the tradition of Newton's physics, it must jettison all talk and concern for the alleged constituents of inner mental life—thoughts, feelings, will, consciousness—and focus strictly on external behavior. Only external behavior could be observed and therefore studied with rigorous empirical experiments:

> I believe we can write a psychology, define it as [the science of behavior], and never go back upon our definition: never use the terms consciousness, mental states, mind, content, introspectively verifiable, imagery, and the like. . . . Psychology, as the behaviorist views it, is a purely objective, experimental branch of natural science which needs introspection as little as do the sciences of chemistry and physics. . . . It can dispense with consciousness in a psychological sense.[79]

The rise of behaviorism suppressed research into the relevance of evolution for psychology. The rationale was this: Just before the rise of behaviorism, the apparent relevance of evolution for psychology consisted in evolution's role in explaining the presence and strength of human instincts. But psychological research—by those studying animal behavior and by the early behaviorists— suggested that instinct played little role in human behavior. Instead, it seemed that human behavior could be molded in any number of ways toward any number of ends.[80] Because the basic impulses of human psychology were seen as something bestowed and shaped by external factors after birth, the role of evolution in making the mind was reduced to insignificance.[81]

Needless to say, psychology was becoming as fragmented and professionalized as the rest of the scholarly world. As we will soon see, this trend greatly influenced how a scientific footing for morality would be pursued later in the century and beyond.

And a Longstanding Anxiety about Moral Skepticism

From the days of Grotius, philosophers and scientists alike recognized the disquieting possibility that, in the end, no moral belief could be established as objectively true; that no moral properties could be demonstrated; and, therefore, that no grounding for any substantive moral belief could be finally justified. The moral disagreement and violence of post-Reformation Europe intensified this fear.[82] As early as the sixteenth century, Michel de Montaigne wondered if moral truth could ever be known. As he put it,

> For what nature had truly ordered for us we would without doubt follow by common consent.... Let them show me just one law of that sort—I'd like to see it.... What am I to make of a virtue that I saw in credit yesterday, that will be discredited tomorrow, and becomes a crime on the other side of the river? What of a truth that is

bounded by these mountains and is falsehood to the
world that lives beyond?[83]

Despite all the optimism of the Enlightenment, none of the val-
iant efforts to find a scientific foundation for morality quelled the
quiet, unsettling, yet growing plausibility of moral skepticism.

Hume is particularly noteworthy here, though perhaps in an
unexpected way. His skepticism about knowledge in general is
well known—indeed, few articulations of skepticism were bolder
or more formidable than Hume's.[84] But his less obvious moral
skepticism is also noteworthy. Even if he didn't *intend* his moral
theory to be skeptical, that is where it points. To see this, begin by
recalling that Hume's view of morality was as scientific as he was
capable of making it: he took a third-person, "outside observer"
stance and attempted to explain moral thought in terms of laws
connecting behavioral traits and feelings, based on careful observations of how human beings feel and act. The view he proposed
marks a strong departure from what we might think of as a
commonsense perspective on moral psychology. Intuitively, we
might think we judge other people as good or bad because we
know something of goodness and evil when we see it, and thus
can correctly classify it when it becomes apparent in another's
action or character. On this commonsense view, the reason we
judge an action evil is that we recognize the evil of the action or
the actor. But in Hume's view, recognition doesn't enter into it,
for there is no good or evil for us to recognize outside of our own
feelings. Rather, an agent judges an action as evil because when
she encounters those sorts of actions they give rise to an unpleasant feeling. Hume's laws of moral psychology connect actions
and traits with special pleasant feelings and therefore call those
traits "good." The same laws connect other behavioral traits with

special unpleasant feelings and therefore call them "evil." His explanation is entirely in terms of a kind of emotional physics between feelings and observed behavior.

This aspect of Hume's moral theory concerned some early readers. Francis Hutcheson read a draft of Hume's treatment of morality in his *Treatise* and criticized him on grounds that his account lacked "a certain Warmth in the Cause of Virtue, which, you think, all good Men wou'd relish, & cou'd not displease amidst abstract Enquirys." Hume actually agreed and offered a vivid metaphor to illustrate why he thought "warmth" was missing: "I must own, this has not happen'd by Chance, . . . There are different ways of examining the Mind as well as the Body. One may consider it either as an Anatomist or as a Painter; either to discover its most secret Springs & Principles or to describe the Grace & Beauty of its Actions." But he thought this separation of perspectives could not be avoided:

> I imagine it impossible to conjoin these two Views. Where you pull off the Skin, & display all the minute Parts, there appears something trivial, even in the noblest Attitudes & most vigorous Actions: Nor can you ever render the Object graceful or engaging but by clothing the Parts again with Skin & Flesh, & presenting only their bare Outside.[85]

Both Hume and Hutcheson recognized that Hume's account fails to capture the nobility and significance that ethical thought and behavior seems to have when considered from the perspective of ordinary experience. So what explains the loss of this warmth when morality is viewed from Hume's scientific perspective? One plausible explanation begins with the shift of perspective from first-person experience of morality to the exterior, third-person view taken by science. The apparent goodness and evil of actions and character that makes sense as the rationale for our moral judgments from our own perspective turns out *not* to be the explanation for our moral judgments on Hume's view.[86] Recognition of real good and evil is an idle wheel in Hume's theory. Hence the genuineness of moral reality is irrelevant. The laws of moral psychology he proposes have no need of that hypothesis.

But Hume doesn't see this as a reason to revise his theory. We don't think the difference between viewing the human body as an artist and as a surgeon gives reason to think either view is invalid. So neither should we dismiss either first-person experience or Hume's dispassionate moral theory—or so Hume seems to say.

Yet his metaphor doesn't seem to work. Even though we feel differently when regarding the human body from the outside than when regarding the interior, we understand that the two are legitimately connected. The bone, muscle, and sinew hold the skin where it should be. They are typically out of sight, but their roles and manner of connection are understood. We don't think we are getting anything *wrong* when admiring the human exterior, even though we know it looks different underneath. The beauty isn't merely an illusion concealing a grisly truth.

The dissonance for Hume's moral theory, however, is of a different category. For he is *denying* that the experience of morality—of the rationality and intelligibility of ethical thought—is as it seems. He's not saying that the way morality seems to us is genuine, but that it is underlain by a cold mechanics that we normally don't see. The dissonance in Hume's moral theory is that, given what he thinks the underlying mechanics are, the way morality seems to us turns out to be *illusory* or *incorrect*. There's no reconciliation between the typical first-person view and his scientific third-person physics of the moral mind.[87] As we show in later chapters, this dissonance—the scientific perspective's inability to preserve the genuineness of ethical experience—lingers unresolved in naturalistic ethical theory down to the present day.

With Darwin, another avenue for skepticism opened. If ethics is simply an adaptation that evolved by natural selection, then we acquire another reason to think it has no compelling justification. Ethics had no being, no ontology beyond whatever our genes and brains and environment generated to keep the social world functioning. Darwinian metaethics thus further weakened the case for an objective foundation for ethics.

By the early twentieth century, other versions of moral skepticism were emerging (though the philosophers' name for these views—"noncognitivism"—unhelpfully obscures that they are in fact a variety of moral skepticism). The logical positivists (including Moritz Schlick, Rudolf Carnap, Otto Neurath, and the young Oxford philosopher A. J. Ayer) argued that if there couldn't be any empirical evidence for or against the truth of a claim, then that claim wasn't actually asserting anything. Since ethical claims are not empirically verifiable (nor are they true or false by definition), the logical positivists judged that moral theory, insofar as such a theory attempted to make moral claims at all, was nonsense.

Emotivism (advocated by such figures as C. L. Stevenson and Ayer) was, in part, spun out of logical positivism, though it had its own distinct take on moral language. In emotivism, the question was simply that if ethical claims are neither true nor false, what are we doing when we engage in ethical discourse? After all, it certainly *seems* like we are making claims—that we are saying things that are either true or false. The emotivist answer was that ethical language was a way of expressing sentiment about various states of affairs, not an assertion of propositions with ethical content.

Thus, an utterance such as "Murder is immoral" would be analyzed by emotivists to really be saying something like, "Boo for murder!" while "It is good to love one's children" might be analyzed as "Hurrah for loving one's children!"

In an environment of growing social and cultural fragmentation, the cynicism of commercial culture, and the failure of science to forge a foundation for morality, moral skepticism, in its varied expressions, would proliferate.

CHAPTER 4

The New Synthesis

A S WE HAVE SEEN, evolutionary approaches to ethics all but died out at the beginning of the twentieth century. Meanwhile, utilitarianism, while still regarded as a viable philosophical theory, existed in numerous variants, each passing different verdicts on what was right and wrong. The utilitarians were left bickering among themselves. As Sidgwick had observed nearly a century before, the idea that empirical study might confirm utilitarianism was not tenable.

The trend away from the science of morality was reinforced by other developments within academic philosophy. The emerging school of logical positivism gave renewed energy to finding scientific foundations for law and morality, yet it concluded that moral claims were not empirically verifiable and were therefore meaningless. Science could explain what human beings were doing when they made moral claims, but the truth or falsity of those claims could never be empirically established and so must be ignored.

The philosophical challenges were formidable on their own terms. The scientific problems were equally daunting. After roughly two hundred years of effort, the science of morality had little to show: no empirical results provided any guidance toward peace among those in conflict, and no mechanics of human moral thought had become clear enough to enable political leaders to engineer flourishing societies. What people regarded as right, good, and virtuous remained a product of religion, contingent cultural currents, and basic intuition. The quest to find a scientific footing for morality languished.

This is not to say that traditional philosophical ethics disappeared. By the second half of the twentieth century, philosophers had begun to theorize in rigorous, systematic ways about right and wrong, good and bad. Several factors accounted for this return.[1] Among the most important was the publication of John Rawls's book *A Theory of Justice*,[2] which showed that philosophers could productively theorize about substantive moral matters. While Rawls's theory was not empirical, it was, in fact, rationalist, meaning that he relied on arguments and methods of reason in order to justify various foundational moral claims. The implications were extensive not only in political theory, where Rawls operated, but in moral psychology. The book had enormous influence in the growth and maturation of cognitive developmentalism, a rationalist stage theory of moral development championed by Rawls's Harvard colleague Lawrence Kohlberg.[3]

In this context, the quest to find a scientific foundation for morality staged a return.

SOCIOBIOLOGY AND ITS DISCONTENTS

The pill that revived the science of morality was found while trying to resolve a longstanding puzzle in evolutionary biology. For much of the twentieth century, it was commonplace among evolutionary theorists that natural selection would favor the survival and reproduction of those individual organisms that behaved selfishly. The question that nagged at evolutionary theorists was how to explain the development of the tendency in some animals to behave altruistically—to act in ways that are not in their individual best interests for survival and reproduction.[4]

Biologist E. O. Wilson published his groundbreaking book *Sociobiology: The New Synthesis* in 1975. In it, he used the research of recent decades showing how social behavior depended on evolutionarily developed genetic factors to suggest an answer to the altruism question: while natural selection favored selfishness in individuals, it favored altruism in groups.[5] Most of Wilson's book concerned the genetic basis of insect social behavior, but in a few places he extended the thesis to human beings. He claimed that "the time has come for ethics to be removed temporarily from the hands of the philosophers and biologicized."[6] Wilson thought the evolutionary explanation for the development of human behavior would prove more fruitful for the study of morality than traditional philosophical ethics had. He predicted that insights from evolutionary biology would combine with those from psychology and neuroscience to provide a new understanding of human emotion, which in turn would explain morality. As he put it,

> Ethical philosophers intuit the deontological canons of morality by consulting the emotive centers of their own hypothalamic-limbic system. This is also true of the developmentalists, even when they are being their most severely objective. Only by interpreting the activity of the emotive centers as a biological adaptation can the meaning of the canons be deciphered.[7]

In essence, Wilson accepted Hume's understanding that morality is emotional mechanics, but Wilson wanted to dig deeper to uncover the neural causes of these emotions, their chemistry and biology. Deeper still, he wanted to investigate how and why evolutionary processes produced human moral emotions. Wilson thus merged the Enlightenment strategy of sentimentalism with an increased focus on evolutionary origins while bringing together several scientific subdisciplines, all in the name of the scientific study of morality. In short, he thought Hume would come galloping back into the discussion, join forces with Darwin, and together with new scientific theories and technologies, they would illuminate morality as never before.

And Hume *would* come back, though not right away, for sociobiology was immediately controversial. There were two main lines of objection. First, certain high-profile biologists—most notably Richard Lewontin and Stephen J. Gould—argued that sociobiology was a form of biological determinism and thus an impediment to a liberal social agenda aimed at overcoming racial and economic inequalities. After all, if human social behavior was seen to be a product of our genes, then initiatives to change society would be rejected as impossible in principle. Second, some objected that sociobiology—quite apart from its unsavory political implications—was just bad science. Philip Kitcher, for one, among a fusillade of objections, pointed out the dangers of describing animal behavior in anthropomorphic language. Doing so fosters "a largely unexamined collection of unsubstantiated hypotheses, latent in our linguistic usage, that allow us to pass freely from conclusions about nonhuman animals to conclusions about ourselves."[8]

Sociobiology was hobbled by such criticism but not destroyed. While its practitioners had initially overreached and been rebuked for it, still it had usefully brought different research programs across disciplines into a coherent whole. It reoriented the quest to find a scientific foundation for morality by bringing together hypotheses and evidence from different fields to explain how complex social behaviors and psychological impulses could have evolved. The notions of kin selection and group selection, for example, made a way for understanding how altruistic behavior—suitably understood—could have developed via natural selection. While human societies do display diversity in their moral beliefs and practices, sociobiology showed that there are, nevertheless, elements of morality that seem universally present and so could have evolutionary explanations.

Elements of a New Synthesis

By the 1990s, sociobiology had reemerged as "evolutionary psychology."[9] Its reconstitution coincided with a new interest in the psychology of emotions—what Jonathan Haidt calls the "affective revolution"[10]—and with developments in the study of the social behavior of nonhuman primates. Primatologists, for example, noted the rudiments of altruism and empathy—the psychological building blocks of morality—in the behavior of chimpanzees and bonobos. These observations provided fodder for psychologists seeking an evolutionary account of human moral impulses.[11] If nonhuman primates showed even hints of moral impulses present in humans, then the human impulses could be explained as an intensification of those hints over time.

The scientific study of the brain, which Wilson had gestured to in the 1970s, also began to proliferate in the 1980s and '90s with the development of neuroimaging technology,[12] which provided extraordinary detail about neurological activity. Naturally, such studies were considered of first importance in revealing how the mind works.[13]

These previously independent strands merged into the more or less unified scientific study of the mind that we might call the "new synthesis," as prophesied in the subtitle of Wilson's *Sociobiology*.[14] It is as much a "science of the mind" as any Enlightenment *philosophe* might hope for, though fueled by the novelties of Darwinian evolution and unparalleled technological access. As we've noted, the new synthesis assumes a broadly Humean picture of moral psychology and takes the evolutionary development of the mind as the paradigm for understanding its function. This marks the return of two of the Enlightenment strategies for a science of morality.

The usually unmentioned but typically assumed big-picture view here is philosophical naturalism, or just "naturalism" for short. Recall that naturalism is the view that nature is all there is, a slogan made more precise as the idea that whatever exists can in principle be completely described in the language of science. On this sort of view, it obviously makes good sense to study morality scientifically. So Hume's psychological account of morality—an account leaving out undetectable moral features in favor of purely psychological phenomena—would seem promising. It makes good sense that the function and origin of our moral psychology would be explained by Darwinian evolutionary pressures. Likewise, it makes sense to understand the goals of human action in more scientific terms, hence (as we will soon explain) the prevalence of a utilitarian-like attitude in practical matters.

So the new synthesis fits hand-in-glove with the view that morality has its roots in moral psychology and thus can be ultimately explained by the scientific study of the mind. This new-synthesis view of morality has four basic elements: (1) a Humean mind-focused sentimentalism, (2) a Darwinian evolutionary account of why the mind has the traits it does, (3) a human interest–based utilitarianism about morality, all embedded within (4) a strident naturalism committed to empirical study of the world. Let's examine these elements in greater depth, with attention to how their proponents articulate them.

First, Hume's approach to the mind—that it can be studied scientifically, via observation and the systematization of data. This turns the study of moral judgment into something more like psychology than traditional moral inquiry. It also emphasizes the general categories Hume believed were necessary to adequately describe morality. These are the moral emotions. In short, against the rationalist bias that had long dominated moral psychology, there is now the recognition that "moral reasoning [is] often a servant of moral emotions."[15] Do people believe in human rights because such rights actually exist, or do they feel revulsion and sympathy when they read accounts of torture and then invent a story about universal human rights to justify their feelings? The Humean view is the latter: that what moral philosophers are really doing is "consulting the emotive centers" of their brains and then fabricating justifications for their feelings.[16]

There is, of course, much debate about the degree to which the emotional and rational are dominant.[17] Yet overall, the Humean element of the new moral synthesis holds that "moral reason works with and through the emotions";[18] that "moral intuitions are one kind of emotional consciousness."[19] As Owen Flanagan put it, this view "has truth on its side" and also "has science on its side."[20] Innovations in neuroscience are important because they help us answer basic questions about morality, namely why you might be concerned with the goals and well-being of people besides yourself. In the new moral science, it turns out that people "have special kinds of neural populations that make concern for others very natural."[21]

Second, in a fuller flowering of Darwin's original insights, evolutionary psychology explanations are taken to show how the mental features necessary for human morality emerged. That is to say, *why* the mental phenomenon of morality is the way it is has an evolutionary explanation. The origins of morality have the same type of explanation as any other biological phenomenon. Primatologist and evolutionary psychologist Frans de Waal says,

> The moral law is not imposed from above or derived from well-reasoned principles; rather, it arises from ingrained values that have been there since the beginning of time. The most fundamental one derives from the survival value of group life.[22]

The turn from moral philosophy to moral psychology is reinforced by the realization that our moral psychology is a product of our evolutionary past. Human moral disagreement, for example, is a result of the conflict between the emotion-based moral psychology adapted for our ancestors' hunter-gatherer lifestyle and the very different demands of modern life. We tend to react and judge the way we do, morally speaking, because of the survival value those reactions and judgments had during the late stages of our evolution. Our gut moral impulses, then, are the direct product of our evolutionary history.[23]

Indeed, all of our basic moral impulses exist because of their survival value. As Patricia Churchland puts it,

In all animals, neural circuitry grounds self-caring and well-being. . . . Lacking the motivation for self-preservation, an animal will neither long survive, nor likely reproduce. . . . Why do we, and other social mammals, care for others? This much we know: on average, such behavior must, either directly or indirectly, serve the fitness of the animals involved.[24]

Owen Flanagan echoes the point:

Darwin is a Humean. . . . To the extent that we are egoists, we are egoists with fellow-feeling. We care about the weal and woe of, at least, some others. Second . . . morality is not "something altogether new on the face of the earth." It is not an invention *de novo*. *Homo sapiens*, presumably like their extinct social ancestors, as well as certain closely related species, such as chimps and bonobos, possess instincts and emotions that are "proto-moral," by which I mean that we possess the germs, at least, of the virtues of sympathy, compassion, fidelity, and courage. . . . We didn't create the relevant instincts and emotions. Natural selection did.[25]

In sum, the building blocks of our moral sentiments, our moral impulses, are what they are because of the role they played in our evolutionary development.

Third, a new utilitarianism is understood as the best means to guide human action. There are, of course, differences between the utilitarianism of Bentham and Mill and that of the new moral scientists.[26] Mill at least seems to have thought that certain states of reality—human happiness, for instance—had real moral value. As a rule, within the new moral synthesis, which course of action is "better" or "best" is a moral issue only insofar it is a practical one.

As Patricia Churchland puts it, relative to human values, "some solutions to social problems are better than others, *as a matter of fact*; relative to these values, practical policy decisions can be negotiated."[27] In this way, these new moral scientists see the question of the moral life as a sort of engineering problem. We begin, Flanagan argues,

> by admitting we are looking for norms, values, and practices that are the best, where "the best" is almost always "the best for such and such purpose or purposes." The analogies are from engineering and the crafts. Given that we need/want bridges that work and shoes that don't leak, we cull from all the possible ways of accomplishing these things the ones that work best.[28]

"Such normative inquiry," he goes on, "has very precise analogies in the sciences such as engineering and botany. Given certain ends, how best can we achieve them?"[29] For some, ethics is, "in its origins at least, a social technology":[30]

> Once ethics is viewed as a social technology, directed at particular functions, recognizable facts about how those functions can be better served can be adduced in inferences justifying ethical novelties.[31]

To these ends, Paul Thagard contends that we should "adopt a normative procedure that empirically evaluates the extent to which different practices achieve the goals of knowledge and morality."[32] The normative procedure, he argues, depicts how descriptive evidence can help establish prescriptive norms:

> 1. Identify a domain of practices. 2. Identify candidate norms for these practices. 3. Identify the appropriate goals of the practices in the given domain. 4. Evaluate the extent to which different practices accomplish the relevant goals. 5. Adopt as domain norms those practices that best accomplish the relevant goals.[33]

Of course, as for the other new moral scientists, Thagard's "goals of morality" are made up of human interests and desires, for they address people's "objective vital needs without which they would be harmed in their ability to function as human beings."[34] Here again, these values are not indicative of any interest-independent or mind-independent value. The standard of human well-being here is nothing beyond the wholly natural realm of human interests and preferences.[35] Which courses of action are "better" is, for the new moral scientists, ultimately a question of utility, and utility is best approached empirically and instrumentally as a sort of practical human engineering project. The ethical project can be understood "as a series of ventures in dynamic consequentialism."[36]

Fourth, the new synthesis draws self-consciously on a very old stream of "philosophical naturalism" in its view of the mind. The central idea of naturalism is roughly that the only things that exist are those that can be described in the language of science.[37] Progenitors of this view date back before the Enlightenment in the work of Bacon, Hobbes, and Spinoza, and later in Hume, Darwin, Watson, and Ryle, among many, many others. One strain of naturalism seeks to provide empirical explanations for all of reality by fitting it into a domain of interacting physical particles.[38]

This would render purely metaphysical or transcendent accounts of reality not only unnecessary but unthinkable.[39] The idea of the mind as an independent, immaterial thing that interacts with the body has become, for naturalists, increasingly implausible[40]—a perspective that is only reinforced by the "cognitive revolution" and brain scan technology, which superficially seem to permit empirical explanation of the mind.

In one sense, there is nothing new in the naturalism that underlies the efforts of the new synthesis. In another sense, there is a different tone in the discourse advocating it: while it operates as a presupposition to this line of thinking, it is a presupposition that rarely goes without saying. There is a polemical edge to today's naturalism and, as we will see, a certain radicalization that brooks no qualification or dissent.[41]

We call the project of the new synthesis view of morality the "new moral science" and those accepting or contributing to it "new moral scientists," even though not all of them are trained as scientists. While we here describe the thought of Joshua Greene, Jonathan Haidt, Patricia Churchland, Owen Flanagan, Alex Rosenberg, and Paul Thagard, plenty of others are in this camp: Michael Ruse, Marc Hauser, Fiery Cushman, Christopher Boehm, Tamler Sommers, Stephen G. Morris, and Mark Johnson, to name a few.[42] The new moral scientists are like many of the thinkers in early and middle modernity: they pursue questions that take them out of the narrow confines of specific disciplines and into work that draws from a wide variety of sources in both the sciences and humanities. We call them "scientists" because each of them ultimately looks to science either to substantially inform moral theory or to provide us with a plan to achieve our goals.

THE LONGSTANDING PUZZLE

The streams of thought that constitute the new synthesis view of morality, then, are anything but new. They have been around for generations, even centuries. What makes the new synthesis view *new* is, first, the particular combination of research concepts that makes up its approach: Hume's mind-focused sentimentalism, Darwin's evolutionary account of the mind, Bentham's and Mill's utilitarianism all embedded within a naturalism committed to empirical study of the mind. In the new synthesis, all efforts to understand morality tend to square with these elements. This configuration of views now appears to be more or less standard across disciplines studying the mind and is common in scientific or scientifically informed approaches to morality. The second source of novelty is the advanced technology offering deeper and more detailed empirical observation of the brain.

The cultural logic of the new moral science plays out in many ways, not least of which is its approach to the longstanding problem of social and moral conflict. The narrative that has emerged is framed roughly as follows: in the past, societies were governed by various "deontological" moral beliefs—those that placed duties to friends, family, or god(s) over those to humanity in general. These, it is argued, were necessary to help humans cope with life in small tribes. For this reason, they have long shaped our intuitive judgments. The problem is that, in modern society, those intuitions, beliefs, and judgments often provoke or justify conflict.

As Joshua Greene put it, people of different societies and cultures "fight not because they are immoral but because they view life . . . from very different moral perspectives." It isn't a matter of "who is right" and "who is wrong" in these conflicts. For Greene, focusing on whether there is moral truth isn't what matters most for practical purposes.[43] The goal is rather to figure out how best to secure objective improvement in satisfying people's interests. The question is, "How can we resolve the problem of conflict?"

> What really matters is whether we have direct, reliable, non-question-begging access to the moral truth—a clear path through the morass [of competing moral values]— not whether moral truth exists.... I'm confident that we don't have this kind of access.... Once we've resigned ourselves to working with the morass, the question of moral truth loses its practical importance.... Resigned to the morass, we've no choice but to capitalize on the values we share and seek our common currency there.[44]

Thus, he argues that we can never solve the problem of moral disagreement on its own terms because moral truth, *even if it exists*, is unknowable. We must therefore transcend the narrow-minded instruction of our intuitive moral judgments, weaken the grip of our felt duties to friends and family, and rely on reason-based judgments. What we should do, in short, is based on what appears to have the best consequences.[45]

Conflict rooted in moral difference is not just one case study among many other moral challenges. It is a thread of concern that has for centuries animated the quest to place morality on a scientific footing. The puzzle of moral difference and the conflict it often fosters continue to perplex and exasperate us all—certainly as much as it ever has. How do we live together with our deep differences? The difficulty this problem continues to generate is equaled only by the hope in the philosophical and scientific communities that it can be successfully addressed through science.

PART III
The Quest Thus Far

CHAPTER 5

What Has *Science Found?*

FOR ALL of the hope that has animated the quest to establish a scientific foundation for morality, there has also been a great deal of actual scientific effort given to the task. So, what are the significant results thus far?

This question is of paramount importance. Finding direct and substantial discoveries about moral issues would all but lay the issue to rest. Failure to find anything plausibly informative would bode poorly for the quest, given how much time it has been afoot, and especially given the huge technological advances in our investigative tools.

But figuring out what science has found is not straightforward. We begin by addressing the issue of how we are to understand the idea of *science* at work in the quest. We then turn to the equally important issue of the multiple relationships scientific evidence can bear to moral matters. Finally, we review the most noteworthy scientific findings that either allege or are alleged by others to tell us something important about morality. That is, we examine the most promising and straightforward scientific attempts to tell us something interesting about right and wrong, about how we should live.

SCIENCE AND ITS BOUNDARIES

For the sake of clarity, it is important to pose the prior question: what exactly do we mean by "science"? We don't want to define science so broadly that anything counts—that wouldn't be interesting and wouldn't capture the force that a real science of morality would offer. Nor do we want to define it so narrowly that genuine, disagreement-resolving empirical studies are excluded. So we need to discuss where to draw the line.

There is a popular misconception that the nature of science is a settled matter—that there is a clear and well-defined consensus among scholars and laypeople about what it is and how it works. In fact, for all its prestige, we actually don't have an accepted criterion for what is and isn't science.[1]

The broadest definitions tend to equate science with rational inquiry itself. Sam Harris, for one, comes close to taking this view, in which science is little more than secular rationality.[2] As Steven Pinker put it, "Anyone who engages in secular reason is a kind of honorary scientist." This view "define[s] science in the broadest term of relying on logic and evidence rather than on dogma or authority or subjective feeling." Yet this approach, he acknowledges, is "not the way most people use the word science."[3]

A more robust definition of science attempts to tie it to the observation of empirical data, mathematical representability, testability, and falsifiability of hypotheses. These may or may not be shared by all rational inquiry. As Paul Thagard says,

> Science uses explanations that are mechanistic and
> mathematical, observations that are systematic and
> made by instruments more powerful than human

senses, and experiments that generate evidence acutely relevant to the choice of explanatory hypotheses.[4]

For our part, when we speak of science, we mean something close to Thagard's description. The reason is that the only approach to science that could provide a science of morality worthy of the name is one closely tied to what is empirically observable. For no other sense of "science" holds any promise of resolving moral questions in the way the harder sciences have resolved the questions they've addressed. Not to put too fine a point on it, broader rational inquiry has not given us a consensus on moral questions, as twenty-five hundred years of philosophical debates amply demonstrate.

Even with this view in mind, it is clear that not every scientific claim is equally well-supported, nor is every claim equally amenable to empirical verification or refutation. As in all sciences, data in a science of morality must be interpreted, and the theories used in interpretation can fit the data more or less closely and can include more or less philosophically speculative content. Of course, the less closely a theory fits empirical data—and the more speculative content it incorporates—the less well supported it is empirically. Theories with loose fit and extra bells and whistles can still be the product of science, but they won't be afforded the same credence as those theories that are close-fitting and elegant.

Finally, and most germane to our argument, science might tell us something about the moral realm in different ways. Consider three: The most interesting way—always the highest aspiration of those who have sought a scientific foundation for morality—would be if science could settle longstanding moral questions. Call this level of scientific results "Level One." Level One results would provide specific moral commands or claims about what is genuinely valuable. They would demonstrate with empirical confidence what, in fact, is good and bad, right and wrong, or how we should live.

"Level Two" findings, while falling short of demonstrating some moral doctrine, would still give evidence for or against some moral claim or theory. For instance, if there was empirical evidence that virtue theories of ethics were false, but the evidence fell short of settling that this or that moral claim was correct, that would constitute a Level Two result.

"Level Three" findings would provide scientifically based descriptions of, say, the origins of morality, or the specific way our capacity for moral judgment is physically embodied in our neural architecture, or whether human beings tend to behave in ways we consider moral. Evidence for these sorts of views doesn't tell us anything about the content of morality—what is right and wrong—but they speak to the human capacity for morality and in that sense are interesting.

Overall, it is fair to note that few scientists actually make Level One claims. In *The Moral Landscape*, Sam Harris claims that science will eventually offer Level One demonstrations, but he doesn't claim to have any such evidence himself or to know of any.[5] As we will see, even Level Two claims are rare. In fact, nearly all of the actual science attempting to deal with morality lands at Level Three findings.

What Does Science Show Us about Morality?

While it would be impossible to review *all* of the literature advancing scientific descriptions of moral phenomena, we can give a few representative illustrations from various fields of research. Our choices are not random but represent some of the most significant and promising cases from each field. In an effort to be objective and charitable, we identified the most-cited articles attempting any sort of scientific exploration of morality in the most prominent generalist science journals and in the top specialist journals for evolutionary biology, evolutionary psychology, neuroscience, social psychology, and primatology.[6] We also attempted to identify the most-cited books presenting scientific approaches to moral theory. We then focused on those works that appeared to achieve the best combination of influence and seriousness.[7] Overall, the empirical findings fell into two broad categories relating to (1) other-regarding behavior and (2) how morally relevant social decisions are made.

EMPATHY, ALTRUISM, AND OTHER-REGARDING BEHAVIOR AS ILLUMINATED BY . . .

Evolutionary Biology

If natural selection favors biologically "selfish" organisms, how could organisms exist that apparently behave altruistically? The primary contributions of evolutionary biology to issues of morality arise from efforts to answer this question—to find ways organisms could have evolved to support the survival of other organisms at a cost to their own survival prospects. This is an old question, of course, but several hypotheses about how this could happen have emerged from evolutionary biology. The leading candidates are "kin selection" and "reciprocal altruism."

To get a sense of how these evolutionary strategies work, it helps to think of the usual case, where selfishness succeeds. The observable traits and macroscopic features of organisms largely stem from an underlying genetic basis—this much is familiar to most. You have brown eyes because your parents had certain genes and passed them on to you in a certain combination. Natural selection, over time, favors those genes that tend to be passed on more successfully than those that don't. Usually, this happens when genes give rise to features that contribute to the organism's being more likely to survive and reproduce than others. This is the typical understanding of how genes contribute to their own reproduction.

But there is another way. Some genes might cause their possessor to behave in ways that actually hurt its own chances of survival, but which benefit other organisms *carrying those same genes*. Such genes would have an overall advantage over other genes, even while the survival chances of the individual organisms possessing these genes might be compromised. This idea is called "inclusive fitness," and it was discovered as a possible evolutionary strategy in a progressive, somewhat piecemeal fashion in the mid-twentieth century.[8]

One way inclusive fitness can proceed is by way of kin selection. In many species of organisms, for instance, parents clearly tend to care for their offspring. This care is altruistic because it comes at a cost to the parents, diminishing their own survival prospects. But because the genetic basis for this care is propagated in the offspring, which typically outnumber the caring parents, such altruism contributes to these genes' success. In kin selection, this altruism is bad for those who are altruistic but good for the genes that make them altruistic. This is why altruism in individual organisms can make evolutionary sense.

Altruism could also arise from a certain kind of reciprocal relationship. Under certain conditions, cooperation between two organisms could give survival advantages to both.[9] Even though each particular action costs an organism while giving benefits to another, over time each organism receives benefits back from the cooperating partner, and the cooperating pair ultimately benefit more overall than those that don't cooperate in this way.

Kin selection and reciprocal altruism, then, are ways evolution could account for some instances of organisms behaving in ways that aren't purely selfish from a biological standpoint. A great deal of work in evolutionary biology is aimed at discovering observable cases that plausibly illustrate these dynamics.

But how far does the empirical evidence actually take us? Biological altruism is a well-established observation: some organisms do sometimes behave in ways that benefit others at a cost to themselves.[10] Beyond this, what evidence shows that kin selection or reciprocal altruism is the mechanism bringing this about? While evolutionary biologists have come up with plausible mechanisms that potentially could explain how biological altruism arose, it's a further step to claim that this is how it actually happened. Here the evidence is promising, but less substantial.

Many scientists now accept that kin selection offers the best explanation for eusociality in insects.[11] A community is "eusocial" when it cooperates in caring for its young and when it includes multiple generations living at the same time as well as participating members who do not reproduce. Bees are the usual example here. Many female worker bees care for the many offspring of one queen, the offspring mature to join the worker bee ranks, and these workers do not reproduce—that is left to the queen. Worker bees behave in ways that benefit their sisters rather than pursuing their own reproduction. As the biologist Jerry Coyne explained,

> Kin selection was an important explanation for the evolution of eusociality. Some think it's because of

the peculiar "haplodiploid" nature of inheritance in Hymenoptera, whereby the male who fertilizes the queen is haploid (has only a single set of chromosomes), and the fertile queen is diploid, with the normal two sets. In such a case, the female workers are more related to their sisters than to their own offspring, which may help them evolve the tendency to stop having their own offspring and produce more sisters; i.e., become sterile and help the queen raise her brood.[12]

In short, kin selection may be the best explanation here: even though worker bees act against their own survival prospects, they do so in a social arrangement that, overall, reproduces their genes very effectively. This would make the genes for biological altruism in bees more fit than their competitors.[13]

This leaves open the question of how biological altruism might help explain *moral* altruism in humans, or whether these two altruisms are even interestingly connected. In other words, while evolutionary biology has given good evidence that biological altruism exists and arose via certain evolutionary mechanisms, is biological altruism a moral idea? Is it the same thing as moral altruism—behavior that is morally good? Or do these homonymous terms mask a crucial difference? We discuss this issue in depth in chapter 7. For now we note that kin selection does seem to be a good explanation for some forms of *biological* altruism.

Evolutionary Psychology

As we've noted before, the aim of evolutionary psychology is to study human behavior and human nature as the products of evolved psychological mechanisms, in order to discover and describe aspects of how and why human minds work.[14] This approach is not without its detractors. A common criticism of evolutionary psychological explanations has been that they are "just so" stories—that they provide no scientific evidence for their account, but at best present one possible way the feature in question could have given an adaptive advantage to the organism that has it. For any given feature of an organism, there are many ways it *could have* come to exist—many combinations of contingent historical and environmental factors could explain its emergence. Hence, coming up with one possible explanation tells us nothing about how the trait actually came about.

But in some cases, even providing a plausible account is worthwhile. One issue that has been all but intractable is how human moral thought and behavior could have arisen via natural selection. What evolutionary benefit does it provide? Here even a "just so" story would amount to an "it's not impossible" story, thereby representing an advance.

The work in evolutionary psychology most relevant to moral inquiry centers on attempts to tell this kind of story. For instance, Frans de Waal defends the hypothesis that empathy—the "ability to be affected by and share the emotional state of another"— provides the psychological motivation for altruistic behavior in animals, in particular some of the great apes.[15] Though trained as a primatologist, de Waal engages a wide literature in making an evolutionary psychological proposal. In the conclusion of his influential article "Putting the Altruism Back into Altruism: The Evolution of Empathy," he states,

> Empathy could well provide the main motivation making individuals who have exchanged benefits in the past to continue doing so in the future. Instead of assuming learned expectations or calculations about future

benefits, this approach emphasizes a spontaneous altruistic impulse and a mediating role of the emotions.[16]

This could provide a causal explanation for why humans don't merely dispassionately increase each other's survival prospects, as could be explained via biological theories of kin selection or reciprocal altruism. We aren't robots seeking to maximize our life spans. We feel things deeply and often are motivated to ethical action from these feelings. Hence the interest in de Waal's hypothesis: it seeks to connect altruistic behavior with emotional responsiveness to others' emotional states. This could perhaps explain why acting on empathetic feelings in human beings has positive consequences for the human species. (It is also worth noting the appeal here to "associationist" notions of psychology, as championed by Hume and Bentham—that our thoughts can be explained in terms of the "physics" between mental states, instead of by rationally directed thought.)

De Waal's argument begins with the observation that certain animals "display distress in response to perceived distress" in others of their species.[17] For such distress to count as sympathy it must include "feelings of sorrow or concern for a distressed or needy other."[18] This sort of sympathy has been observed: "Perhaps the best-documented example of sympathetic concern is consolation, defined as reassurance provided by an uninvolved bystander to one of the combatants in a previous aggressive incident."[19] To move from sympathy to empathy requires that the animal not merely experience concern for the other but also share its emotional state. This, too, has been observed: "A major manifestation of empathic perspective-taking is so-called targeted helping, which is help fine-tuned to another's specific situation and goals. The literature on primate behavior leaves little doubt about the existence of targeted helping, particularly in apes."[20] De Waal and a colleague propose a specific mechanism for *how* the mind engages in empathy. The capacity for empathy is made possible by "a mechanism that provides an observer (the subject) with access to the subjective state of another (the object) through the subject's own neural and bodily representations."[21] This mechanism somehow enables the animals doing the observing to identify with and experience the emotions and felt needs of an observed animal. (Recall again Hume's "strings of sympathy.")

While there appears to be no clear, observational evidence of animals engaging in empathy-based altruism,[22] plenty of evidence exists that they behave altruistically.[23] The empathy mechanism could help explain how and why some animals do behave altruistically—in the *biological* sense. This explanatory utility supports de Waal's account of the motivational bases for altruism: "This is, in fact, the beauty of the empathy-altruism connection: the mechanism works so well because it gives individuals an emotional stake in the welfare of others."[24]

As with the findings of evolutionary biology, whether de Waal's theory might help us move beyond *biological* altruism to *moral* altruism must wait until chapter 7. But the empirical case for the evolutionary development of biological altruism via empathy seems plausible.

Primatology

Some of the most important findings from primatology for the study of morality involve the development of the human sense of justice. Capuchin monkeys, chimpanzees, and domesticated dogs show negative reactions when treated worse than social partners of their same species. That is, if these animals are given lesser rewards than their social partners, they quit engaging with the activity/reward test setup.[25]

For example, in their highly cited 2003 article "Monkeys Reject Unequal Pay," Sarah Brosnan and Frans de Waal report that Capuchin monkeys sometimes will refuse a reward for performing a task if other Capuchins are given a bigger reward for the same task.[26] The interest of this research is that it might shed some light on how both human cooperation and a sense of fairness evolved. Not because humans descended from Capuchins; we didn't. But the Capuchins' moral or moral-like behavior is less developed than ours, and learning how and why proto-moral behaviors and sensitivities might have developed reveals possibilities for how our own moral behaviors might have evolved as well.

More specifically, the idea behind this study is that cooperation might make evolutionary sense only if enough cooperating individuals benefit from it, and if the benefits are distributed fairly equally. If so, we might expect individuals not to cooperate if the benefits are unequal. The Brosnan and de Waal study supports this prediction: Capuchin monkeys do appear to display "inequity aversion."

Moral Cognition Revealed By . . .

Neuroscience

Perhaps the most exciting scientific research on human morality of late has come from neuroscience. Numerous experiments made possible by the invention of brain scan technology have begun to show what happens at the neural level when human beings consider moral issues. A typical experiment involves doing functional magnetic resonance imaging (fMRI) scans on subjects while they consider versions of the famous trolley cases. The two best-known (and most commonly used) versions of the trolley case are the switch case and the footbridge case.

In the switch case, a runaway train is bearing down on five people trapped on the track, and it will kill them if no one intervenes. But there is a lever one can pull that will switch the train to another track, where only one person is trapped. The test subjects are asked whether it is morally acceptable to switch the train to the second track, thereby killing one but saving the five. The subjects undergo an fMRI scan of their brains while they are considering their answers.

The footbridge case begins much like the switch case: a runaway train is bearing down on five people trapped on the track, and it will kill them if no one intervenes. But here is the difference: instead of a lever that one can pull to divert the train, in this case there is a large man standing on a footbridge directly over the track, between the oncoming train and the five trapped people. If the man is pushed off the bridge, he will fall onto the track in front of the train, slowing it just enough to prevent it from killing the five. Of course, the large man will die from the collision. The test subjects are asked whether it is morally permissible to push the man off the bridge, killing one but saving five. Again, their brains are scanned via fMRI while they ponder this question.

For what it's worth, about two-thirds of test subjects say that it is morally acceptable to switch the train using the lever, while only one-third say it is acceptable to push the man from the bridge to stop the train. But what's interesting is that the fMRI scans showed that people use *different parts of their brains* for the two cases. When considering the switch case question, people had increased neural activity in their brain's dorsolateral prefrontal cortex (DLPFC)—approximately the front fifth of the brain. Neural activity in this region is associated with consciously controlled, impersonal, calculative thought. On the other hand, when considering the footbridge case, people show increased activity in their ventromedial prefrontal cortex (VMPFC). This region is located in the middle of the front bottom part of the brain and is associated with automatic, unconscious, reactive emotional thought.

From this finding, Joshua Greene has developed what he calls the "dual-process" theory of moral judgment:

> It's a dual-process theory because it posits distinct, and sometimes competing, automatic and controlled responses. . . . In response to the switch case, we consciously apply a utilitarian decision rule using our DLP-FCs. . . . The harmful action in the switch case does not elicit much of an emotional response. As a result, we tend to give utilitarian responses, favoring hitting the switch to maximize the number of lives saved. . . . In response to the footbridge case, we also apply the utilitarian decision rule using our DLPFC. But here, for whatever reason, the harmful action does trigger a (relatively) strong emotional response, enabled by the ventromedial prefrontal cortex (VMPFC). As a result, most people judge that the action is wrong, while understanding that this judgment flies in the face of the utilitarian cost-benefit analysis.[27]

To help clarify the difference between the judgments characteristic of these two processes, note that the cold, deliberative, calculative thoughts associated with determination of which decision saves more lives arises from the DLPFC, while the visceral repulsion associated with the thought of pushing the man to his death arises from the VMPFC. Greene compares our two processes with two settings on a camera: emotion-influenced judgment in the VMPFC is quick, automatic, and out of our conscious control, like the camera's automatic setting. The impersonal, calculative judgment influenced by the DLPFC is slow and consciously controlled, like the manual setting on a camera.[28]

If Greene is right, neuroscience shows us that our moral thought is dual process. But that's not all. Greene notes that there seems to be a connection between manual-mode thinking and ethical utilitarianism on the one hand and between automatic-mode thinking and rights- and duties-based ethical views. As he puts it,

> Characteristically deontological judgments are preferentially supported by automatic emotional responses, while characteristically consequentialist judgments are preferentially supported by conscious reasoning and allied processes of cognitive control.[29]

Here Greene uses the philosophical jargon of "consequentialism" and "deontology," but these are proxies for utilitarianism on the one hand and for ethical views emphasizing the importance of rights and duties on the other.[30]

In sum, Greene and his colleagues have shown via fMRI studies that moral judgment is *dual process*. That is, it characteristically draws on intuitive emotional judgments as well as calculative, effortful judgments. Furthermore, his studies have shown a correlation between particular areas of the brain and whether a moral judgment is emotional and automatic or calculative and "manual."

This much seems to be empirically well supported (though we discuss it in greater depth in chapter 6), and it amounts to several Level Three findings on morality. It tells us something new about how our moral thinking relates to brain activity. These sorts of observations, if ultimately confirmed, could lead to fruitful collaboration between scientific and philosophical accounts of moral phenomena, as well as concrete application to human capacity for moral thought. For instance, knowing more about *where* certain kinds of moral thinking occur could prove helpful in diagnosing and treating people exhibiting impaired or pathological moral cognition. The value in Greene's work here is obvious.

And in Social Psychology

A series of studies conducted in the 1970s revealed that even very slight differences in a situation strongly affect people's behavior. For instance, in what was perhaps the most well-publicized study, researchers found that people who had found a dime in a payphone were twenty-two times more likely to help a women who had dropped a stack of papers than those who had not found one.[31] In another study, people were five times more likely to help an injured man who had dropped a stack of books if noise levels were normal than if a loud lawnmower was running nearby.[32]

What makes these data interesting and relevant to the study of morality is the relationship between the data and a category of moral theories known as virtue theories.

Virtue theories differ from consequentialist and deontological ethical theories in that they take as the central moral phenomenon not good consequences or absolute rights and duties but the character traits exemplified by virtuous people. According to virtue theories, what you should do in any given situation is what a person with virtuous character would do. Virtue theory thus emphasizes character over actions. But what is a character trait? In the literature, a character trait is defined as a relatively stable disposition to act in a certain way:

> If someone possesses the virtue of courage, for example, she is expected to *consistently* behave courageously across the full range of situations where it is ethically appropriate to do so, despite the presence of inducements to behave otherwise.[33]

What studies purport to show, however, is that people do not display these sorts of traits. Rather, they seem to act in response to different situations. By these lights, virtue theories appear not to provide a good account of ethical goodness.

We want to highlight the apparent significance of these findings. If correct, they would seem to provide empirical evidence against a moral theory, and as such constitute Level Two results. We discuss these findings in greater depth in chapter 7, and point out a grave difficulty for its status as Level Two. Nevertheless, this is one of the few scientific studies explicitly alleged to provide direct guidance concerning the correctness (or incorrectness) of moral theories.

The Psychology of "Moral Foundations Theory"

The studies described above are sometimes presented as Level Two findings, but except for the social psychological criticism of virtue theory, they are more accurately placed in Level Three: they tell us about the human (and animal) capacity for morality but say little about the content of specific moral claims or the truth of moral theories. Assertions of Level Two findings are rare.

Even so, there isn't space for a comprehensive overview, so we explain here one of the more significant and interesting theories purporting to address a Level Three question: moral foundations theory (MFT). Proposed by Jonathan Haidt and others, MFT is an attempt to explain the psychological bases of similarity and difference in moral judgments across cultures, whether these bases are unified or plural, and how they might have been evolutionarily adaptive—that is, why these bases might have given early humans an evolutionary advantage over competitors who lacked them.[34]

Haidt and his colleagues have attempted to reduce the profusion of moral values and systems across cultures to five or six basic elements. These elements are thought to be the innate, psychological building blocks that then are modified and expressed in different ways in different cultures. Haidt and his colleagues draw an analogy to taste receptors. Human beings innately possess five basic kinds of taste receptor: sweet, salty, sour, bitter, and savory. But there is nonetheless great variation across cultures in which of these tastes a person prefers, which kinds are emphasized in the food he or she eats, and which combinations are accepted as palatable. Some cultures focus more on sour foods, whereas others nearly exclude the sour. Something similar, Haidt suggests, holds for human moral sensibilities. MFT is an attempt to identify the "best candidates for being the innate and universal 'taste receptors' upon which the world's many cultures construct their moral cuisines."[35] So far, Haidt has identified six basic moral "tastes" or modules: (1) care/harm, (2) fairness/cheating, (3) loyalty/betrayal, (4) authority/subversion, (5) sanctity/degradation, and (6) liberty/oppression. The claim is that for any given human being, moral values will be driven by the culturally shaped manifestation of some or all of these basic sensibilities. Despite positing a universal psychological basis for morality, MFT can accommodate moral disagreement. Haidt illustrates this with the following question:

> Should parents and teachers be allowed to spank children for disobedience? On the left side of the political spectrum, spanking typically triggers judgments of cruelty and oppression. On the right, it is sometimes linked to judgments about proper enforcement of rules, particularly rules about respect for parents and teachers. So even if we all share the same small set of cognitive modules, we can hook actions up to modules in so many ways that we can build conflicting moral matrices on the same small set of foundations.[36]

MFT is a young theory that is still being refined. The sixth moral taste receptor—liberty/oppression—was added in just the last couple of years, well after the original four or five.[37] Other theoretical refinements are doubtless in the works. For instance, one of the current requirements for a value to count as a moral module is that it be generated by an intuitive, "automatic" cognitive process, rather than be arrived at via reflection, calculation, or reasoning more generally.[38] But it seems plausible that this constraint has little to do with whether a human value is psychologically basic or not. Instead it reflects an independent concern regarding the value's relationship to reason.

Nevertheless, in outline if not in all its details, MFT has been supported by many kinds of evidence: self-report surveys, implicit measures from reaction-time tests, neuroscientific studies independent of self-report, and text analysis to locate terms tied to moral foundations in various cultures' literary works.[39] Perhaps the best-known support for MFT has been its facility in explaining the psychological basis of the moral disagreements between liberals and conservatives in US politics.[40]

If MFT is correct, part of its value will be in the discovery that human beings share basic moral sensibilities even across wide moral disagreements. Jesse Graham and Haidt speak to this point by quoting Isaiah Berlin: "The difference this makes is that if a man pursues one of these values, I, who do not, am able to understand why he pursues it or what it would be like, in his circumstances, for me to be induced to pursue it. Hence the possibility of human understanding."[41]

MFT is an illuminating and interesting account of human moral psychology. But it tells us little if anything about which morality is the correct one. Not that it intends to; Graham and his coauthors emphasize that MFT is a descriptive project that says nothing about what is right or good but is merely "trying to analyze an important aspect of human social life."[42]

GRAND AMBITIONS, MODEST RESULTS

After five hundred years of scientific inquiry into the nature of morality, the most noteworthy scientific findings at best achieve Level Three status. Moral foundations theory describes basic and important moral emotions; research in evolutionary biology has uncovered promising mechanisms that illumine how other-regarding behavior could have evolved. It's hard to deny these findings' value for answering descriptive questions about the nature and origin of the building blocks of morality. Perhaps such findings will play a helpful role in some broader case regarding a specific moral question. This shouldn't be surprising: at some level, the ethical and the empirical must connect.

But as far as we can tell, there are no scientific findings that present claims of either Level One or Level Two status. Even the renewed energies of the new synthesis have provided no clear empirical support for any moral theory, let alone for any claim about what is right and wrong, good or evil, or how we should live.

CHAPTER 6

The Proclivity to Overreach

THERE IS, of course, much more scientific research on morality than we can cover in our brief review. What is most relevant for our argument is that even the most highly cited and most highly regarded studies only lead us, at best, to Level Three findings.

Yet there is a notable tendency to overreach—to suggest that certain studies offer Level Two or even Level One findings that reveal aspects of an authoritative morality or adjudicate between competing theories of morality. At times the rhetoric can be extravagant. As moral psychologist Fiery Cushman put it in *New Scientist*,

> As we come to a scientific understanding of morality, society is not going to descend into anarchy. Instead, we may be able to shape our moral thinking towards nobler ends. Which norms of fairness foster economic prosperity? What are the appropriate limits on assisting a patient's end-of-life decisions? By recognizing morality as a property of the mind, we gain a magical power of control over its future.[1]

The question is, is Cushman's confidence in a scientific under-standing of morality well-founded?

Philosophical and Methodological Limitations

The boldness of the scientific claims often conceals substantial philosophical and methodological weaknesses. As one example, let us consider again Joshua Greene's work.

In his book *Moral Tribes* and in his paper in the elite philosoph-ical journal *Ethics*, Greene has argued that the data from social neuroscience and evolutionary biology undermine our intuitive moral judgments and give us reasons to become utilitarians. This bold thesis has invited much comment and criticism.[2] But even his more empirically conservative claims face difficulties.

Greene argues, first, that his neuroscientific experiments empirically support the dual-process model of moral judgment, and second that utilitarian and deontological judgments correlate with the two processes. He presents these as universal claims about human moral thought. But the experimental subjects whose brain activity is taken to be representative of all human moral thought are in fact just a handful of college students at elite, northeastern American universities. Greene's most highly cited study is based on just nine undergraduate students, and the second most highly cited study looked at forty-one Princeton undergrads.[3] These samples are not only small, they are very homogenous in age, social status, and education. They are likely also homogenous in their subjects' religious, moral, and other cultural backgrounds. Yet the moral impulses of these few Ivy League students are meant to tell us about the nature of moral thought for all of humanity: for elderly female Muslim subsistence farmers in northern India, indigenous hunter-gatherers in Papua New Guinea, and ambitious young atheistic businessmen in Shanghai. We have no good reason to think Greene has told us much about the moral thought of humanity, when all he has studied is the moral thought of a few Princeton students, especially given the likely influence of factors such as age, culture, and class on moral thought.

But even limiting our focus to the handful of college students he tested, it is still not clear that Greene's studies show what he thinks they show. For instance, one of his more highly publicized claims is that effortful, calculating, "manual" moral judgments in favor of, say, switching the track to save several lives are *utilitarian*. This alleged connection between moral judgments of this sort and the well-known moral theory of utilitarianism has spawned a minor research industry.[4] Greene concludes from this alleged connection between neuroscience and moral philosophy that

> Moral psychology is not something that occasionally intrudes into the abstract realm of moral philosophy. Moral philosophy is a manifestation of moral psychology. Moral philosophies are, once again, just the intellectual tips of much bigger and deeper psychological and biological icebergs. Once you've understood this, your whole view of morality changes. Figure and ground reverse, and you see competing moral philosophies not just as points in an abstract philosophical space but as the predictable products of our dual-process brains.[5]

But there are several difficulties with Greene's thesis.

Consider just one of his basic claims: that he has empirically demonstrated some general connection between manual-mode moral judgments and utilitarianism. For one, the alleged connection is mostly drawn from a narrow range of experimental cases, namely the trolley-type cases described in chapter 5. Since the realm of moral judgment is obviously much wider than trolley cases, this is not a strong empirical foundation for imagining that utilitarian judgments correlate with deliberative, manual-mode thinking in general.[6] Absent studies examining a much broader range of moral scenarios, there isn't much at all we can conclude in general about the relationship between deliberative thought and utilitarianism.

But even in the narrow range of the trolley cases, we cannot move so easily from the empirical fact that deliberative, manual mode thinking plays a causal role in generating characteristically utilitarian judgments in trolley-style cases to the claim that deliberative thought reflects the utilitarian character of those judgments. Instead, for all Greene's studies show, it might be that the utilitarian options in the trolley cases are just *less intuitively correct* from a commonsense perspective. The increased deliberation and manual-mode processing may merely be a result of the unintuitive nature of these positions—"Is it really OK to directly cause the death of one person in order to save five others?" The test subject has to puzzle through the case to see whether he or she can accept the utilitarian answer, since it's unusual and difficult. This would explain the additional time and effort we see in the tests. On the other hand, the deontological response in the footbridge case, for instance, is eminently intuitive—pushing people off bridges is just not done. It doesn't take long to recognize this. As Guy Kahane observed,

> The apparent tie between process and content [deliberative goes with utilitarian, automatic goes with deon-

tological, etc.] is really just an artifact of the kinds of
scenarios that researchers have studied, reflecting
nothing very interesting about utilitarian and deontological judgments.[7]

That this alternative is equally well supported by Greene's
studies is problem enough for his attempt to link reason and
utilitarianism, but Kahane and colleagues have also performed
studies corroborating this alternative hypothesis. Their evidence
suggests that it is intuitiveness/unintuitiveness, rather than deon-
tology/utilitarianism, that typically accounts for deliberation in
moral judgments.[8]

Still, there is much value in Greene's findings. For one thing,
prior to his studies, we didn't know that calculative and emotional judgments tended to correlate with activity in distinct
regions of the brain, or which regions these might be. But what is
the value of Greene's findings for understanding right and wrong,
or how we should live? It is unclear that they have much. After
all, we already knew that moral judgment often has an emotional
component. We already knew that consequentialist judgments
are often less emotional (think of the stereotype of the bloodless
utilitarian bureaucrat). We already knew that some moral judgments are intuitive, not evidently based on conscious reasoning.
(Hence the long history of intuition-based ethics. Sidgwick has
a lengthy discussion of this in *The Methods of Ethics*, published
150 years ago.) We already knew that some moral judgments are
based on conscious reasoning (calculating and weighing the values of competing actions, considering arguments, etc.). So, insofar as Greene's empirical evidence supports these claims, it just
amounts to new, empirical support for things we already knew.
This isn't without value, since it's better to have more rather than

less evidence for our beliefs. But in the end, his theoretical ambitions far outstrip his observations.[9]

Another example of this sort of overreaching comes from primatology. As we reviewed in chapter 5, Frans de Waal argues that some nonhuman primates feel sympathy—and even empathy—for each other, and that this could explain why they help each other. For de Waal, sympathy consists of "feelings of sorrow or concern for a distressed or needy other," and empathy is the "ability to be affected by and share the emotional state of another."[10] Sympathy and empathy in nonhuman primates may then provide an evolutionary roadmap for the development of human altruism.

De Waal is right that human morality often involves feelings of sympathy and empathy, and the actions that these feelings motivate. He may also be right that the development of this sort of altruistic behavior in nonhuman primates tells us something about how our own capacity for altruism developed.

But de Waal has grander ambitions for his research here. In a longer work devoted to the implications of his "sympathy to altruism" argument for human morality, de Waal concludes, "At some point [in human evolutionary history], sympathy for others became a goal in and of itself: the centerpiece of human morality."[11] This is the key link in his project of showing how human morality evolved, as witnessed in his title *Primates and Philosophers: How Morality Evolved*.

However, there is still a significant gap between behavior being altruistic—even in the so-called *moral* sense—and the behavior being *moral*. Acting out of sympathy or empathy for another isn't enough. To appreciate this gap, consider the following case.

Suppose you help a neighbor with her rent payment because you know she is having trouble making ends meet, and you've perceived her emotional anguish over the coming shortfall. Your capacity for empathy enables you to understand what she's feel-ing—the anxiety, dejection, shame, and other emotions—and even to feel a little bit of this yourself. Your capacity for sympathy enables you to feel concern for her. For the psychological mech-anism proposed by de Waal and his colleagues, this basis is suffi-cient to explain helping behavior in certain nonhuman primates.

But is helping your neighbor with her rent merely on the basis of your sympathy for and empathy with her enough for your action to be *moral*? And if it is morally altruistic, is that enough to actually make it the *right thing to do*?

To see the problem, let's add to the scenario. Suppose you also know that your neighbor is a heroin dealer whose business activi-ties are endangering the lives of your family and your other neigh-bors. But because you empathize and sympathize with her in her predicament, you decide to cover her rent anyway. Of course, by doing so you enable her to stay in the heroin business. Though your helping was motivated by your empathy with and sympathy for your neighbor, it is not clear you have done the right thing, morally speaking. The problem is, at least in part, that in acting this way, you have been inappropriately oblivious to the moral context. Yes, you helped someone, and out of concern for them and their pain, but given the facts that you knew, it seems you should not have done so. And the situation is no different if your neighbor is not a drug dealer at all but a person of spotless charac-ter. If you are helping merely out of reactive concern or as a reac-tion to perceived pain, you may help those who ought not to be helped; you may help those who should be hindered. Empathy and sympathy are not enough.

Certainly, de Waal has shown us a plausible account of the development and psychological functioning of certain tools we draw upon in moral judgment and action. Empathy and sympathy often form part of the basis for our moral behavior. What remains unclear, however, is how his theory of these capacities tells us anything about the nature of prescriptive morality, or about how to live. As we've explained, while empathy and sympathy sometimes play a role in motivating actions that turn out to be moral, they can also play a role in those that are *immoral*. You can act sympathetically yet immorally. With this in mind, calling sympathy the "centerpiece of human morality" is, at best, quite a stretch.

Consider another illustration, this one from social psychology. Here is Steven Pinker in the *New York Times*, reflecting on the work of Jonathan Haidt:

> People don't generally engage in moral reasoning, Haidt argues, but moral rationalization: they begin with the conclusion, coughed up by an unconscious emotion, and then work backward to a plausible justification. . . . The science of the moral sense also alerts us to ways in which our psychological makeup can get in the way of our arriving at the most defensible moral conclusions. The moral sense, we are learning, is as vulnerable to illusions as the other senses. It is apt to confuse morality per se with purity, status and conformity. . . . In the worst cases, the thoughtlessness of our brute intuitions can be celebrated as a virtue.[12]

Pinker then claims that the sense of repugnance—"shuddering"— many feel at the prospect of biomedical manipulation of human life is just such an irrational, merely emotional impulse:

> There are, of course, good reasons to regulate human cloning, but the shudder test is not one of them. People

have shuddered at all kinds of morally irrelevant viola-
tions of purity in their culture: touching an untouch-
able, drinking from the same water fountain as a Negro,
allowing Jewish blood to mix with Aryan blood, toler-
ating sodomy between consenting men.[13]

Certainly, some have felt repulsion toward things that turned
out upon examination to be morally fine, but this doesn't license
rejecting all moral judgment based on repulsion. Instead, it may
be that we merely need to calibrate our responses of repulsion.
This sort of calibration is routine in the life of any reflective
agent and is required for good reasoning generally. After all, some
have thought they had rational arguments for this or that, but it
turned out upon examination that they didn't—they made a mistake. Obviously, the right response isn't to condemn outright our
attempts to rely upon reason, but rather to refine it. So if sometimes getting it wrong doesn't undermine our attempts to use
reason, why should it undermine our attempts to use repulsion?
Pinker operates with a double standard here.

And Pinker's position is even weaker than we've let on. Suppose, for the sake of argument, we agree that repulsion-based
moral judgment is misguided. This claim is not supported by anything like empirical demonstration. No scientific finding shows
or even suggests that repulsion to human cloning is misguided.
Pinker might have philosophical arguments, but so do his oppo-
nents. Science has done nothing to clear up rational disagreement
over the "wisdom of repugnance."[14]

Pop Science—
the Case of the "Moral Molecule"

Perhaps the paradigm case of pop science overreaching is the story told about the role of the chemical oxytocin in human cooperation—the so-called moral molecule.[15]

The story begins with a short paper in *Nature* in 2005, in which a team of five scientists described their finding that spritzing oxytocin into the noses of test subjects correlated with increased trust in interpersonal interactions.[16] The paper noted that part of the interest of the study was that it seemed to shed some light on the mostly unknown biological basis of trust in human beings. Its significance was not lost on other scientists interested in the neurobiology of human social behavior, and soon the paper had thousands of citations.[17]

But what really pushed this study—and the broader issues of science and morality it may pertain to—into the spotlight were the efforts of one of the five original authors. Paul J. Zak, a neuroeconomist at Claremont Graduate Center, began pitching this research to a much broader audience, kicking off with a popular TED talk and moving on to multiple web articles, culminating with his 2012 book, *The Moral Molecule: The Source of Love and Prosperity*.[18] He has since published a second book applying his ideas to the business world, titled *Trust Factor: The Science of Creating High-Performance Companies*. This ascension was accompanied by some antics that were no doubt intended to make his scientific offerings more accessible and memorable. For instance, in his TED talk, Zak repeatedly referred to himself as "Dr. Love" and prescribed "eight hugs a day" in order to elicit a release of oxytocin (which he claims correlates with increased happiness).[19]

But it is Zak's more sober claims that arouse real interest.

Throughout *The Moral Molecule*, he regales the reader with stories illustrating situations in which the release of oxytocin in the brain provides the essential explanatory element in moral behavior.

The basic social experiment that underlies his case is called "The Trust Game," and it works like this: each of two subjects— call them A and B—is given a sum of money. Neither subject's identity is revealed to the other at any point in the experiment. A is given the option of giving any amount of his money to B, and the test organizers explain to both parties that they will triple whatever A gives to B, at which time B will have the option to give back to A any amount he or she wishes—perhaps as a reward or in gratitude for A sharing in the first place, though this is neither required nor encouraged. Traditional economic theory would say that it is in both A's and B's interest to just keep whatever they receive. It is in their best interests not to give any of it away. But behavioral economists have found that both A and B routinely share with the other subject,[20] with the result that each walks away in a better financial position than if they had chosen not to share.[21]

What Zak observed is that those who had trusted more and shared had higher levels of oxytocin, and those who shared more tended to come out of the game with more money than those who shared less. He tried to control for personality differences, couldn't find any, and so concluded that the oxytocin is the key difference-maker here. Zak's takeaway? People whose brains release more oxytocin are more trusting and wind up benefiting more in Trust Game scenarios.

Zak goes on to posit oxytocin as the underlying factor explaining "natural sympathy" among human beings. Along the way he notes certain complicating factors that prevent oxytocin release from reliably establishing trust in human relationships, such as the inhibiting effects of testosterone, stress, experiences of abuse, or genetic abnormalities. Eventually he presents his central theory incorporating oxytocin into human morality:

> Human beings can be both good and bad, but in stable and safe circumstances, oxytocin makes us mostly good. Oxytocin generates the empathy that drives moral behavior, which inspires trust, which causes the release of more oxytocin, which creates more empathy. This is the behavioral feedback loop we call the virtuous cycle.[22]

Zak describes various ways in which empathy drives moral behavior as it prompts us to take the perspective of, and feel something of, the experience of another. He does briefly note that there is more to moral behavior than merely acting out of oxytocin-influenced brain states, pointing out that we sometimes refrain from empathizing in cases where we believe others have brought their predicament upon themselves and so deserve the hardship they're facing.[23]

In fact, as we explained previously, empathy is *never* by itself sufficient to explain moral behavior. But instead of focusing on Zak's philosophical weaknesses, here we want to look at his overreaching: his attempts to extend and amplify scientific results far beyond what they show. In the introduction of *The Moral Molecule*, Zak writes,

> Am I actually saying that a single molecule—and, by the way, a chemical substance that scientists like me can manipulate in the lab—accounts for why some people give freely of themselves and others are coldhearted bas-

tards, why some people cheat and steal and others you can trust with your life, why some husbands are more faithful than others, and by the way, why women tend to be more generous—and nicer—than men?

In a word, yes.[24]

Zak leaves little doubt that oxytocin is the central explanatory factor in human morality:

> After centuries of speculation about human nature, human behavior, and how we decide what is the right thing to do, here at last we have some news we can use— solid empirical evidence that illuminates the mechanism at the heart of the moral guidance system. As any engineer will tell you, understanding the basic mechanism is the first step towards improving a system's output. . . . Given that humans can be both rational and irrational, ruthlessly depraved and immensely kind, shamefully self-interested as well as completely selfless, what *specifically* determines which aspect of our nature will be expressed when? When do we trust and when do we remain wary? When do we give of ourselves and when do we hold back? The answer lies in the release of oxytocin.[25]

In the span of seven years, Zak went from writing that oxytocin merely increases trust in certain interpersonal interaction to stating that oxytocin:

> ▸ Is the "mechanism at the heart of the moral guidance system."

- ▶ Explains the varying degrees of unfaithfulness and generosity.
- ▶ "Specifically determines" the nature and timing of human moral behavior.

There has never been any doubt that moral thought is linked to behavior by *some* physical means. After all, thought often leads to action. Perhaps oxytocin figures in this causal process. But where does it figure—especially with respect to moral judgment and character? Does oxytocin cause or influence moral judgment, or does moral judgment cause or influence oxytocin release? Does the content of the moral judgment matter, that is, does whether you judge something good or bad influence how much oxytocin is released? Or does the amount of oxytocin released influence whether you judge something good or bad? Is having virtuous character a product of oxytocin release, or can virtuous character motivate good moral judgment and action apart from, or despite, oxytocin release?

In order to legitimately claim that oxytocin is the major factor explaining morality, Zak would need to answer these sorts of questions. But he makes little effort to do so. Without being able to show how oxytocin figures in the broader explanatory framework of moral thought and action, he is in the position of someone trying to tell us that what explains drunk driving accidents is the presence of alcohol in drivers' blood. Certainly this is an important, even necessary factor. But focusing only on blood chemistry neglects many other explanatory elements, such as human responsibility in choosing to drive drunk or in choosing to become drunk, or the cultural, genetic, and psychological factors that bear on such decisions. With this broader framework in mind, imagine the significance of a book that explains excessive levels of blood alcohol as the cause of drunk driving accidents. This would not be an enlightening thesis.

As a matter of fact, Zak's case for the relevance of oxytocin in explaining morality is even weaker than this. Up to this point, we've taken for granted that his scientific studies show what he has alleged, even if his attempts to situate these results in an explanation of morality have badly overreached. But even the kernels of real science in Zak's project have been near-fatally undermined. Five attempts to replicate the original Trust Game study that jump-started Zak's project have failed to produce similar results.[26] In lay terms, five other studies tried to show what Zak showed in his Trust Games and couldn't do it. Even Ernst Fehr, one of the scientists who authored the "Trust Game" paper with Zak, now says, "What we're left with is a lack of evidence. I agree that we have no robust replications of our original study, and until then, we have to be cautious about the claim that oxytocin causes trust."[27] The news gets worse: not only has the basic science of the "moral molecule" not been replicated, but additional research has suggested that under some conditions oxytocin promotes aggression and defensiveness, emotions directly opposed to the cuddly ones Zak describes.[28]

In sum, Zak's "moral molecule" project promised to identify the central mechanism of human love, morality, and prosperity, but what he delivered is little more than a slick misrepresentation of modest results that have yet to meet basic scientific standards for replication.

A BLURRED BOUNDARY

The new moral science has generated a great deal of intellectual excitement around the potential of neuroscience, evolutionary biology, and moral psychology to provide empirical discoveries about the nature and functioning of morality. From our vantage point, the most reliable findings are rather modest. At the same time, it is not surprising to find that some of the rhetoric surrounding the promise of the new moral science overreaches and makes the findings sound more robust than the evidence warrants.

This overreaching depends on obscuring the distinction between "Is" and "Ought," between description and prescription. By fudging that line, one may give the impression that practical moral implications emanate from the science; that a special moral authority derives from scientific expertise. This tendency is pervasive, even among the brightest lights of the new moral science.

In an exchange in the *New York Review of Books*, philosopher Tamsin Shaw criticized Pinker, Haidt, and a few other new moral scientists on just this point. Shaw wrote that the blurred boundary

> serves, whether intentionally or unintentionally in each case, to generate the false impression that scientific expertise has important weight in making moral judgments, and that behavioral science can provide us with a guide to moral development. In circumstances in which psychologists are being called upon by their own colleagues, among others, to address very serious moral questions, they have a special responsibility to be clear about any claims they appear to make to moral authority. [I question] whether they have fulfilled this responsibility.[29]

Greene, Haidt, and Pinker will be quick to point out that they explicitly disclaim any link between their descriptive findings and their normative implications. For example, in his most comprehensive work on morality, *The Righteous Mind*, Haidt writes,

> Philosophers typically distinguish between descriptive definitions of morality (which simply describe what people happen to think is moral) and normative definitions (which specify what is really and truly right, regardless of what anyone thinks). So far in this book I have been entirely descriptive.[30]

Likewise, while attempting to explain the role of the moral sense in the decline of violence, Pinker states, "The starting point is to distinguish morality per se, a topic in philosophy (in particular, normative ethics), from the human moral sense, a topic in psychology."[31] Greene also denies "that science proves that utilitarianism is the moral truth."[32]

But Haidt offers this disclaimer only once in *The Righteous Mind*, and only on page 271 of a book of about 350 pages. Pinker makes his statement on page 623 of a 700-page book—and only *after* presenting about 100 pages of scientific research on morality as a central part of his larger ethical claim that the world is improving. Pinker continues to blur the boundary between descriptive and prescriptive morality in his book *Enlightenment Now* when he asserts that "the scientific facts militate toward a defensible morality, namely principles that maximize the flourishing of humans . . ."[33] Greene himself immediately follows his disclaimer by arguing "that utilitarianism becomes uniquely attractive once our moral thinking has been *objectively improved* by a scientific understanding of morality. . . . [We can] use twenty-first-century science to vindicate nineteenth-century moral philosophy against its twentieth-century critics."[34]

The boundaries become even blurrier when Haidt makes definitive moral recommendations based on his descriptive findings.[35] It does not clarify the situation that he holds an endowed chair in ethical leadership and is the founder and director of Ethicalsystems.org, a nonprofit organization that, on the basis of scientific research, advises businesses on cheating and honesty, conflict of interest, corruption, fairness, ethical leadership, whistle-blowing, and the ethical culture of organizations.[36] He's not alone in using his skills for business consulting.[37] "Science can make us better," one scholar declares, because it allows us to shape our moral thinking toward nobler ends.[38]

The tendency to smudge lines is not confined to academics but is woven into the institutional fabric of the funding organizations. The mission of the Templeton Foundation, for example, includes the "hope for advancing human progress through breakthrough discoveries." Noble ends, to be sure, but normative at their core.

The line between description and prescription at times is a matter of open debate, as it was among the marquee scientists, philosophers, and journalists gathered at the Edge conference on "The New Science of Morality." Here they discussed "whether [scientific] work should be seen as merely descriptive, or whether it should also be a tool for evaluating religions and moral systems and deciding which were more and less legitimate."[39]

In popular culture, the lines are actively crossed. The positive psychology movement, for example, draws deliberately and deeply from the new moral science to advance the moral goods of "positivity." The moral goal of improving happiness, well-being, flourishing, and self-esteem suffuses the larger movement, and as Martin Seligman puts it, "It is grounded in careful science."[40]

Our point is not to disparage any of these scholars, or the institutions that give them a platform. Rather, we want to recognize that the discourse is inherently problematic, made even more so by its brightest proponents. It is easy to imagine that people with less expertise in the subject may get the impression that the scientific findings deliver substantive moral guidance. Even among the most circumspect, the boundaries between description and prescription are technically maintained, but blurred in practice. There is plausible deniability, but it is thin to say the least.

CHAPTER 7

Intractable Challenges

T HE BEST RECENT science has provided insight into the nonevaluative elements of morality and offered suggestive possibilities about its claims on human experience. But it has given us nothing remotely close to an empirical foundation for morality—nothing close to an "ought" from an "is."

The preceding two chapters, which focused on the "science" part of the science of morality, looked at the limitations of this inquiry within the practice of science itself. But there are also problems with the "morality" part of the science of morality. The issue is how to *understand* morality for scientific purposes—a difficult question, and one that science itself can't answer.

A genuine science of morality—a science capable of giving us an "ought" from an "is" and thus capable of adjudicating moral differences—must meet at least two challenges. First, it must meet the challenge of definition. That is, it must make clear that the phenomenon it describes really is morality and not merely something that vaguely resembles, approximates, or accompanies morality. Otherwise it will be open to the charge that it isn't really an account of morality at all. It will fail to provide the intellectual consensus necessary for the development of a body of scientific knowledge.

Second, it must meet the challenge of demonstration. For a science of morality to have a chance of resolving moral disagreement, it must have a way of adequately demonstrating the truth of its claims. This demonstration must be the sort of empirical procedure that could change the minds of those with different moral commitments.

The Challenge of Definition

Science requires clarity and consensus about the phenomenon under study. Biology would not be much of a science if biologists could never agree on which things were cells, physics would not be much of a science if physicists could never agree on the properties that define light or gravity, and chemistry would not be much of a science if chemists constantly wrangled over which elements make up the periodic table. Even if they could conjure agreement, they would still not be very good sciences if they failed to define their subjects of study in ways that were adequate to the phenomena they wished to understand. In the same way, it is essential to a science of morality that morality be conceptualized in a way that fits the reality—that convincingly conceptualizes morality and other moral terms in a way that closely, even if not perfectly, describes what they are.

Suppose, for example, we claimed to be able to scientifically demonstrate what is morally right and wrong. Your ears might prick up with interest. We then explain how we're defining key moral terms. We define a "morally right action" as "any action done on a Tuesday" and a morally wrong action as "any action not done on a Tuesday." This permits an easy demonstration of whether an action is morally right or wrong—at least according to our theory. We just figure out whether the action was performed on a Tuesday, and we have our answer. But you would rightly find this approach wanting, because our theory doesn't actually demonstrate anything about morality. We can demonstrate whether certain actions were performed on a Tuesday, but that tells us nothing about their moral status. Being done on a Tuesday isn't, in the end, a good definition of morality. It's not just that we can make arguments about why it is irrelevant to morality. This definition is so far away from what most people think of as morality so as to have no chance of being recognized as morality. Thus it cannot provide the conceptual clarity and consensus necessary for a science of morality.

Scientific theories of morality are, of course, more sophisticated. Still, they often fail to meet the challenge of definition because their definitions of moral terms include items that don't belong (such as merely pragmatic goals—as we'll soon discuss) or exclude things that do belong (such as rights or duties). But they can also fail by using moral terms that are clearly defined but don't fit the context in which they are used. The term "morality" has several legitimate meanings, and which is used has tremendous implications for what a scientific theory of morality is really showing us.

A Lexical Range: The Prescriptive, the Descriptive, and the Prudential

Consider three different senses of "morality" and their implications for scientific theories of morality.

First, "morality" can mean the realm of right and wrong, good and bad, whether these are grounded by fundamental moral laws or by the value of particular things and states of affairs. This is the sense we intend when we say, for instance, that killing innocent people for fun is morally wrong or that racism is immoral. This morality is *prescriptive*, meaning it is supposed to justifiably guide human action. This is the kind of morality we might describe as "genuine," "real," "prescriptive," or "authoritative."

Second, "morality" can mean what people *think* is right and wrong—the realm of social rules and practices, and the rules or decisions that describe what groups of humans believe constrain certain kinds of behavior and encourage other kinds. This is the sense of morality we intend when we talk about a society's moral code without intending to say anything about whether such a society's codes really are right or wrong. This is also the sense of morality under investigation in the vast majority of scientific work on morality.[1] We might call morality in this sense *descriptive*.

Third, "morality" can mean something practical or instrumental. In this sense, it concerns what one should and shouldn't do, but where the "should" isn't a moral "should" in the lived and prescriptive sense. That is, there's a kind of "ought" that is practical without being ethical. It's the sort of "ought" we mean when we say things like, "Well, if you want to win the lottery, then you ought to buy some tickets." In such cases, we aren't saying that anyone morally ought to buy lottery tickets, but instead just that if someone's goal is to win the lottery, then to achieve it they would have to buy some lottery tickets. This kind of normativity is sometimes called *prudential*.

Recent books on the science of morality are full of various claims about what this or that neuroscientific discovery or evolutionary developmental story tells us about morality. But do such claims mean to tell us something new about what really is moral? Do they say anything about what we should or shouldn't do or what is good or bad, beyond what certain people might happen to think we should do? Or do these claims merely describe certain human practices or impulses that are causally related to this or that neurological property or evolutionary history? That we can't have this or that moral feeling without certain localizable C-fibers firing? Which of these is meant matters a great deal for figuring out what is being shown.

Consider an imaginary case. Suppose you wanted to figure out how best to budget your income, manage your debt, and invest for the future. You pick up a few books with promising-sounding titles, such as *Homo Thrifticus: The Evolutionary Science of Personal Budgeting* and *Mind over Money: The Neuroscience of Saving.* You begin reading the books with gusto, but before long a troubling worry arises: Are these books using the scientific investigation of personal finance to reveal how best to manage your money, or merely to reveal how human beings tend to think, feel, and act with respect to their money? Since you know that many Americans have a lot of debt, you doubt that knowing how humans are inclined to use their money will help your budget. The neural basis for what people actually do seems to offer little guidance on what you *should* do. The same goes for the moral life.

But much of the recent scientific study of morality makes little attempt to clarify which sense of morality is under discussion. Those who advocate a science of morality commonly conflate the various meanings of the word, as though these differences didn't exist or matter.

For instance, Christopher Boehm, in his highly regarded book *Moral Origins: The Evolution of Virtue, Altruism, and Shame,* at no point acknowledges that there is a crucial difference between prescriptive and descriptive senses of morality. This is significant because he claims to offer a theory of the evolution of our moral sense—locating his discussion in the age-old story of Adam and Eve and the fall of humankind. The ideas of moral culpability and conscience in the story of Adam and Eve are clearly prescriptive. But for the remainder of his book, Boehm resolutely uses the descriptive sense, vitiating its relevance for telling us anything informative about the prescriptive sense of morality he opened with.[2] Without a firm grasp on the difference between the senses of morality, readers will naturally but mistakenly assume that Boehm's theory is about prescriptive morality—a morality that concerns right and wrong and how we should be and behave.

Moving between "Ought" and "Is"—
A Case from Neuroscience

Consider also Patricia Churchland's influential book *Braintrust: What Neuroscience Tells Us about Morality*. Initially, it seems clear that Churchland intends to use empirical research to illuminate genuine, prescriptive morality. Early in the book, she describes how she thought philosophy had little to offer in answer to philosophical questions like, "What is it to be fair? How do we know what to count as fair?" While she found it plausible that Aristotle, Hume, and Darwin were right that humans are social by nature, "Without relevant, real data from evolutionary biology, neuroscience, and genetics, I could not see how to tether ideas about 'our nature' to the hard and fast."[3] But by drawing on these fields of study and employing a "philosophical framework consilient with those data, we can now meaningfully approach the question of where values come from."[4] She says her aim was "to explain what is probably true about our social nature, and what that involves in terms of the neural platform for moral behavior."[5] This will be helpful because "a deeper understanding of what it is that makes humans and other animals social, and what it is that disposes us to care about others, may lead to greater understanding of how to cope with social problems."[6]

The impression left on the reader is that philosophy can't answer big moral questions but science can, and that Churchland intends to contribute toward answering these questions. It would seem that she is using moral terms in the genuine, prescriptive sense, but a more careful reading makes it is less clear that she is.

First, notice the shift: she begins with questions that are moral in the prescriptive or lived sense: "What is it to be fair?" and so forth. Philosophy couldn't help her answer these questions; she felt she needed "hard and fast" data. So she turned to the sciences, which give her data on the "neural platform for moral behavior," thereby helping her figure out "where values come from." Churchland thinks that examination of human brains reveals the neurochemical nature of certain impulses that give rise to social behavior—she describes these impulses as "values"—and that we can figure out what is best to do given that we have these values:

> The truth seems to be that the values rooted in the circuitry for caring—for well-being of self, offspring, mates, kin, and others—shape social reasoning about many issues: conflict resolution, keeping the peace, defense, trade, resource distribution, and many other aspects of social life in all its vast richness. Not only do these values and their material basis constrain social problem-solving, they are at the same time facts that give substance to the processes of figuring out what to do—facts such as that our children matter to us, and that we care about their well-being; that we care about our clan. Relative to these values, some solutions to social problems are better than others, as a matter of

fact; relative to these values, practical policy decisions can be negotiated.[7]

But what would knowledge of the neural platform for moral behavior or of where values come from tell us about what it is to be fair? It is hard to tell. The "hard and fast" data that Churchland seeks from science have something to do with morality, but do they illuminate prescriptive morality—what we should do—or just tell us something descriptively about the machinery that supports our ability to consider, make, and follow moral judgments? Churchland offers little help in understanding what the relationship is supposed to be between the hard scientific results she describes and the big, genuinely moral questions that originally animated her research.

There is another confusion. Churchland explicitly claims that morality is real. "It is as real as real can be," she says. "Some social practices are better than others, some institutions are worse than others, and genuine assessments can be made against the standard of how well or poorly they serve human well-being."[8] At first blush, it seems as though she is referring to a prescriptive understanding of morality. At the same time, she claims that morality is part and parcel with human social behavior: "Social behavior and moral behavior appear to be part of the same spectrum of actions, where those actions we consider "moral" involve more serious outcomes than do merely social actions such as bringing a gift to a new mother."[9]

The question, then, is what distinguishes merely social norms and rules from those of morality. What distinguishes the merely prudential from the robustly prescriptive? Churchland finds the difference in the "seriousness" of the elements we call "moral."[10]

But what is this seriousness that makes some social actions

moral? Churchland doesn't say. How then are we supposed to tell whether she's talking about real morality—the realm of right and wrong, good and bad—or just descriptive or prudential morality? Despite her assertion that she's talking about real morality, her explanation of what she takes morality to be completely dodges the question.

This sort of avoidance is unhelpful and misleading, but even if Churchland had straightforwardly explained exactly which sense of morality she was talking about, her account would still face grave difficulties. To see this, recall that she says the "serious" social actions are the "really" moral ones. But either this seriousness makes the actions morally correct or it doesn't, and instead, is some kind of ultimately nonmoral way of distinguishing the actions we take to be moral from those we don't. If she means the first—that moral actions really are prescriptively guided—then we're still left with no sense of how the contingent and manipulable presence of neurochemicals could give rise to legitimately action-guiding sentiments or judgments. How does the release of dopamine or serotonin explain a correct judgment about morality? But if she means the second—that the "seriousness" that makes some social actions moral doesn't bestow genuine action-guiding prescriptivity—then she's no longer addressing the genuinely moral questions she originally claimed to be pursuing. Instead, she would merely be articulating a theory about the neurochemical components of our moral judgment, without telling us anything about these components' connection to our awareness and appreciation of morality.

In the end, Churchland can only assert, without evidence, that values are just those things humans care about. Given that we care about certain things, there are facts about what course of action would best achieve or promote these things. The sense of morality that she thinks science illuminates turns out to be what we've called "prudential"—tied to what we happen to want—rather than a morality that might legitimately claim to show us how to live.

The Case of Altruism

Consider the example of altruism as another real-world example of the challenge of definition. In a recent book, biologist David Sloan Wilson argues that science can demonstrate that altruism exists. Whether this is an interesting claim depends entirely on how Wilson defines altruism. He gets off to a promising start, beginning his book by defining altruism as "a concern for the welfare of others as an end in itself."[11] This definition of altruism puts it at least closer to the prescriptive realm, given, among other reasons, the inclusion of goodness of intention and a Kantian concern for human beings as ends in themselves. If Wilson could give a scientific argument for the existence of this kind of altruism, it would be a major breakthrough.

But two different senses of altruism are at play in his argument—one prescriptive and the other descriptive; one ethically intended and directed and the other biologically elucidated. In biology, altruism means one organism increasing another organism's reproductive fitness at a cost to its own. This stands in contrast to an ethical understanding of altruism, where the idea is framed in terms of one's acting with the intention of benefiting another, without regard for the cost to one's self.[12]

Unfortunately, Wilson abandons ethically relevant understandings of altruism in favor of a more empirically tractable behavioral definition. Toward the end of the book, he makes this redefinition explicit:

> Altruism exists. If by altruism we mean traits that evolve by virtue of benefitting whole groups, despite being selectively disadvantageous within groups, then altruism indubitably exists and accounts for group-level functional organization we see in nature.[13]

This account takes us far from any recognizably prescriptive definition of altruism. Wilson is aware of this but unconcerned. His book, he argues,

> has been critical of some of the ways that altruism is traditionally studied. Altruism is often defined as a particular psychological motive that leads to other-oriented behaviors, which needs to be distinguished from other kinds of motives. Once the existence of altruism hinges on distinctions among motives, it becomes difficult to study because motives are less transparent than actions. . . . But to the degree that different psychological motives result in the same actions, we shouldn't care much about distinguishing among them, any more than we should care about being paid with cash or a check. It's not right to privilege altruism as a psychological motive when other equivalent motives exist.[14]

So Wilson claims that satisfying the biological, behavioristic definition of altruism is all that really matters. As he put it, "It doesn't matter whether he gets paid in cash or by check."

But this is far from evident. Do you only care that your spouse *acts* as though she loves you? That she says complimentary things to you, that she appears to enjoy conversation with you, that she does her share of the household chores, that she contributes income to the family, says the things a loving spouse should say, appears to be sexually attracted to you, and remembers your birthday? What if you discovered that she does all of these things without feeling anything for you—or worse, she does all these things while secretly detesting you? Would Wilson claim that this is just a "cash or check" situation—just so long as she's doing all the observable things she would do if she really did love you, then the underlying motives, intentions, and desires are irrelevant? It is difficult to imagine that he would be indifferent to his own spouse's motives and intentions. Such a relationship would be functional, but loveless—missing precisely the element that makes acts genuinely altruistic.

Wilson's definition of the key moral term in his study ultimately renders his account at best incoherent, and possibly irrelevant to a science of altruism. This is the common outcome for scientific accounts that falter on the challenge of definition.

All-Too-Convenient Definitions of Virtue

Recall from chapter 5 the "situationist" argument from social psychology that claims to undermine virtue theories of ethics. The evidence allegedly demonstrates that people sometimes do not act from stable virtuous character traits. But a key problem with the empirical case against virtue theory is the proposed definition of *character trait* at work.[15] For it isn't at all clear that character traits are as simple as the proponents of these studies have made out.

As Gopal Sreenivasan has clarified, a character trait isn't merely a rigidly stable disposition to act given the presence or absence of this or that factor.[16] Rather, there is reason to think a character trait is a "multitrack" disposition, possession of which means that the agent has the disposition to act in the relevant way, but also is sensitive to competing concerns that might outweigh the decision to act in accordance with the disposition. For instance, the mere fact that someone tells a lie isn't by itself indication that the subject in that case has done something inconsistent with possession of the virtue of honesty (as some studies have assumed).[17] For the subject may have judged that while telling the lie was wrong, it was better to do so than to, say, snitch on a classmate. In such a case, we might well classify the subject as acting in a way consistent with the virtue of honesty, even though he or she for overriding moral reasons did not tell the truth in that instance. As Sreenivasan puts it, this sort of "lying situation would still not be a good behavioural measure of honesty, unless we wanted to define a model of honesty as someone who behaves 'honestly' even when the balance of reasons stands against it."[18] Given that this oversimplified definition of a character trait is often assumed in these social psychological studies, they fall short of showing us anything about the presence or prevalence of such traits.

The mistake here is firmly on the "definition" horn of the dilemma. All the actual science employed here seems straightforwardly legitimate. People did not manifest certain kinds of dispositions in the empirical studies. The problem arises in selecting the correct conception of character trait to judge against the data.

There are potentially interesting consequences for moral theory from these sorts of studies. For instance, *if* one holds a version of virtue theory according to which character traits are as simple and rigid as assumed in (the relevant interpretation of) the studies, *then* the studies appear to provide empirical evidence that traits so imagined are at best quite rare. This would constitute at least a Level Three finding concerning morality.

Definition and Specificity

Questions of method and measurement are critical to science, but they are irrelevant if researchers cannot clearly mark out the object of inquiry in a credible, consistent, and persuasive way. Conceptual clarity is essential, yet it is often missing in the new moral science. In the shell game that results, scholarship often presents itself as addressing questions of prescriptive morality but, through a sleight of hand, puts descriptive and prudential definitions of morality into play in ways that conflate the meanings of the term. The promised prescriptive conclusions somehow disappear.

The absence of conceptual clarity brings another fundamental problem as well, and it bears on what, for scientific purposes, is understood to be moral reality. Invariably, the science of morality is directed toward unearthing and explaining universally shared moral principles. These are ethical generalities that take shape as moral-philosophical abstractions. The evidence used to address this moral reality purports to be species-wide, whether it is drawn from data from neurochemistry, the evolutionary record, or public opinion surveys.

This is fine as far as it goes, but it barely scratches the surface of morality as it exists in the lives of individuals, groups, communities, and nations. While it may be possible to speak of universal moral principles, nearly all of what we actually know and experience of morality exists only in its particularity: in a bewildering array of complex and contradictory moral traditions, stories, and ideals—made all the more complex by race, ethnicity, gender, region, political economy, and history. This empirical complexity arouses little curiosity from the new moral science. It is as if the best way to address empirical difference were to ignore it altogether.

But any intellectual inquiry that disregards empirical specificity fails to meet the most rudimentary requirements of a science. More basic still, without the rigors of an inductive method working upward through empirical complexity, there can be no confidence that the concept of morality used in any given study will resemble morality as it exists in individual and social life. Such inquiry can only produce vague generalities and broad speculations, and not conclusions rooted in scientific rigor.

The challenge of definition, then, is a formidable challenge. But there is another.

The Challenge of Demonstration

Early in the quest to find a scientific foundation for morality, Hugo Grotius thought that postulating rights would help establish a moral basis for laws that "if you rightly consider, are manifest and self-evident, almost after the same Manner as those Things are that we perceive with our outward Senses."[19] The problem, of course, is that rights—if such things exist—are not as evident to the senses as Grotius thought. Joshua Greene explains the problem this way:

> Appeals to "rights" function as an intellectual free pass, a trump card that renders evidence irrelevant. Whatever you and your fellow tribes-people feel, you can always posit the existence of a right that corresponds to your feelings. . . . Rights and duties are the modern moralist's weapons of choice, allowing us to present our feelings as nonnegotiable facts. By appealing to rights, we excuse ourselves from the hard work of providing real, non-question-begging justifications for what we

want. . . . We have, at present, no non-question-begging way to figure out who has which rights.[20]

In short, rights, even if they exist, can't be demonstrated.

The problem isn't that rights are irrelevant to morality or couldn't help explain it. The problem is that insofar as morality is explained in terms of rights, to that degree it can't be empirically demonstrated. Perhaps this isn't a problem for most of us, even for most moral theories. But it is an ineradicable problem for any theory of morality that intends for its claims about morality to be in any way empirically observable or demonstrative.

This is the challenge of demonstration: for a theory of morality to be scientific, it must tie its claims to observable reality strongly enough to demonstrate that it is getting the nature of morality right. Put more sharply: a science of morality must be able to demonstrate empirically that its claims about morality are true.

The Case of Well-Being

The new moral scientists are fond of providing examples to show that science has demonstrated (or can demonstrate) the truth or falsity of certain moral claims. A favorite is the health or medical analogy. For instance, Michael Shermer, author of *The Moral Arc: How Science Leads Humanity Towards Truth, Justice, and Freedom*,[21] writes that

> Historically, we have already been using science to determine such moral values as the best way to structure a polity, an economy, a legal system, and a civil society, in the same way that physicians have developed improved medical science and epidemiologists have

worked to build better public health science. . . . If you agree that it is better that millions of people no longer die of yellow fever and smallpox . . . and many other assaults on the human body, then you have offered your assent that the way something is (diseases such as yellow fever and smallpox kill people) means we ought to prevent it through vaccinations and other medical and public health technologies.[22]

This shows, he believes, how science can determine moral values.

But Shermer's argument doesn't show that science can determine or demonstrate moral values. It merely shows that if being a certain way makes something immoral, and if you can tell that something is that way, then you can tell that it is immoral. This would count as demonstration only if that first assumption—that being a certain way makes something immoral—was itself demonstrable. But it isn't—or at least, no one has yet been able to provide a case where it is. In other words, Shermer's argument must assume at the beginning the values he claims can be demonstrated scientifically.

Now, there's nothing wrong with reasoning as Shermer does in his argument. If you think that having a certain observable property makes a thing good or bad—say, a car's being fast makes it good, or a dog's eating the sofa cushions makes the dog bad—then you are entitled to conclude from observations about the speed of cars and the diet of certain dogs that these things are good or bad. But unless someone already agrees with you about fast cars or sofa-eating dogs, you can't convince her that these things are good or bad just by pointing out that the car is fast or the dog has ingested a cushion. This method may have evidential value for those who share your value assumptions, but it has no evidential value for those who don't. It can't provide agreement via demonstration.

Sam Harris follows much the same line of reasoning. Harris is a bit more circumspect than Shermer in that he recognizes that he's assuming that certain observable properties are tied to certain moral values. Harris puts it this way:

> Science cannot tell us why, scientifically, we should value health. But once we admit that health is the proper concern of medicine, we can then study and promote it through science. . . . I think our concern for well-being is even less in need for justification than our concern for health is. . . . And once we begin thinking seriously about human well-being, we will find that science can resolve specific questions about morality and human values.[23]

Harris makes two assumptions—first, that well-being is a moral good, and second, that we can identify the observable properties of well-being. Yet he doesn't see this as problematic for the scientific status of his argument. After all, he reasons, we make similar assumptions in medicine, but we can all recognize that it is still a science. But Harris still doesn't recognize that these assumptions are fatal to his claim that science can determine moral values. To make the problem more vivid, compare Harris's argument with arguments that share the same logic and structure:

- ▸ Science cannot tell us why, scientifically, we should value the enslavement of Africans. But once we admit that slavery is the proper concern of social science, we can then study and promote it through science. I think our concern for embracing slavery is even less in need for justification than our concern for health is. . . . And once we begin thinking seriously about slavery, we will find that science can resolve specific questions about morality and human values.

- Science cannot tell us why, scientifically, we should value the purging of Jews, gypsies, and the mentally disabled from society. But once we admit that their eradication is the proper concern of social science, we can then study and promote it through science.

- Science cannot tell us why, scientifically, we should value a prohibition on gay marriage. But once we admit that such a prohibition is the proper concern of social science, we can then study and promote it through science.

These parallel arguments are horrific and outlandish to our ears today. Yet they all have surprisingly recent historical precedent. Most tellingly, they rely on the same logic as Harris's. But of course they have little hope of showing that we should approve of slavery, prohibit gay marriage, and eliminate Jews, gypsies, and the mentally disabled. Why? Because these arguments merely *assume* that we should and then recommend the scientific study and promotion of these ends. Harris is doing the same thing. But his response here conceals a sleight of hand.[24] When he applies this argument to health, it seems more powerful than it really is because most of his readers already agree that health is a good that can be illuminated via science. But once we see that the same argument can be applied just as readily to values we do not agree with, it becomes clear that the assumption of a particular value doesn't make it scientifically demonstrable.

Harris's defense is that values are intrinsically tied to science from the outset—for example, the value of truth. So why not the value of health? Isn't health something valuable? And if health is in, then why not the value of well-being? Recall that the question is whether science can demonstrate or determine value. While you can't scientifically demonstrate that truth is valuable, it is true that you have to assume the value of truth to do science. It's hard to see how you could pursue science well without it. That's why it's acceptable to assume the value of truth here. But does the same thing go for the value of health or well-being? Certainly not! As we all know, there can be and, sadly, have been all sorts of scientific pursuits oriented *against* health, well-being, and human flourishing.

In the end, what Harris is proposing is not the scientific determination of values, but that science can show us how to promote something we've already *assumed* is valuable, independent from science. So what he presents isn't really science but is rather science-plus-a-value-assumption, where the value assumption cannot be demonstrated empirically. This is fine so far as it goes: the world should have room for confessional pursuit of answers given assumed starting points. But these "science-plus" approaches must not be confused with the sort of science that has given us chemistry, physics, and biology. They place Harris on the same field of play as anyone who wants to add a controversial, even abhorrent value assumption to science. Harris promises a legitimate, chemistry-worthy science of morality, but all he can muster is a "science-plus," wherein the interesting moral claim— the "plus"—is asserted rather than demonstrated.

The Case of Happiness

The challenge of demonstration plays out differently in the proliferation of studies addressing the phenomenon of happiness or well-being in human life.

First, the idea that happiness could be tightly coupled to morality is historically undeniable. In the Western world, in both ancient and medieval thought, happiness presupposed an objective moral order to which one would conform one's life. For Aristotle, for example, happiness was not the effect of momentary pleasure but the result of cultivation of a life lived virtuously—in his own words, "an activity of the soul in accordance with virtue." For Aquinas, earthly happiness was ephemeral and rather beside the point, but true happiness could come from contemplation of or nearness to the divine, and it always came through strenuous moral rigor and spiritual discipline.

It wasn't until the Enlightenment that anything like the contemporary notion of happiness was given expression. As Louis Antoine de Saint-Just put it in 1794, "Happiness is a new idea." While far from true, his observation called attention to the meaning of happiness as "positive emotion" or "psychological well-being." The idea that it should be pursued as an end in itself *was* new.

There is some variation today. Most define happiness in the tradition of Bentham, as a subjective state of buoyancy, positive feeling, pleasurable emotions, merriment, and the like.[25] Others attempt to define it more broadly in the Aristotelian tradition, as flourishing that includes valuable activity.[26] But even in the latter effort, the virtues are not understood as intrinsic and mind-independent goods but are redefined into functional capacities oriented toward generating positive emotions. The road still leads back to Bentham.

What is taken as "happiness" today is far from a universal feature of the human condition, but rather an artifact of modern history. What is more, the empirical studies that seek to understand happiness reflect even further bias by orienting those studies overwhelmingly toward people who are "Western, educated, industrialized, rich, and democratic" (the acronym "WEIRD"). These groups, which are among the least representative in the world, serve as both the subjects for the experiments and the primary audience for the findings.[27]

The historically and culturally tendentious way in which the new moral science defines happiness is a problem for those making claims about the universality of human nature. It is compounded by a demonstration problem: the empirical part of the "science" of happiness relies on respondents' self-reported levels of subjective positive emotion that specify, along a single metric, how satisfied they are with their lives.[28]

Let us leave aside the dubious idea that happiness is one-dimensional: at least the definition of happiness is clear. The demonstration problem is that there is no objective or natural unit by which we can measure happiness.[29] What people actually mean by happiness or satisfaction is not known, and therefore it cannot be meaningfully compared from one person to another or from one time to another. Certainly, in the post-Enlightenment West, the ethical value of subjective positive feelings seems self-evident. But the support for this belief isn't empirical. As James Pawelski put it, "Perspectives shift dramatically. What is taken for granted about happiness in one cultural context seems foreign in other cultural contexts. . . . Because of these dramatic shifts, we must avoid the mistake of thinking that our current views on happiness necessarily hold true for cultural contexts different than our own."[30]

The challenge of demonstration looms larger still. Suppose, for the sake of argument, that we could measure happiness in a meaningful way. Suppose we could calculate the presence and degree of people's subjective positive feelings, their deep engagement, how meaningful they take their life to be, the positivity of their relationships, and their accomplishment. What would this tell us about morality? By itself, nothing. For these data to have relevance to questions of morality, we would have to know that subjective positive feelings, deep engagement, and the like are ethically good or are worth pursuing. But no empirical technique yet known can uncover this. The criticism here is the same as our criticism of Shermer's argument: he assumes the values he claims to demonstrate empirically. Those surveys may tell us, in a very approximate way, people's levels of positive emotion or life satisfaction, but the ethical relevance of these properties is beyond the scope of the studies. As Martha Nussbaum put it,

> Pleasure is only as good as the thing one takes pleasure in: if one takes pleasure in harming others, that good-feeling emotion is very negative; even if one takes pleasure in shirking one's duty and lazing around, that is also quite negative. If one feels hope, that emotion is good only if it is based on accurate evaluations of the worth of what one hopes for and true beliefs about what is likely.[31]

As we say, there is no way to scientifically demonstrate the goodness or value of the ends to which happiness is directed.

AN INTERNAL BARRIER

Can one derive an "ought" from an "is"? Hume's Law notwithstanding, some continue to try to do so.[32] The question continues to intrigue and confound. But it defies any simple resolution because of the enduring challenges of definition and demonstration.

In Bentham's "felicific calculus," for instance, the basis for morality was found in how much pleasure resulted from an action, but it failed because it couldn't generate agreement that pleasure and the absence of pain were the only valuable things in the moral life. Nor could the utilitarians convince their critics that valuable things formed the only basis for morality—that there were no duties or rights that could trump calculations in terms of valuable consequences.

Across four centuries, approaches to the science of morality can be positioned roughly between two points on a spectrum of possible definitions of morality. On one end are definitions that involve genuine goods—"oughts," duties, rights, and the like—that have some real prescriptive authority over human behavior. At the other end of the spectrum are definitions that permit empirical assessment—definitions consisting of terms that can be observed or measured.

Both the historical record and the leading contemporary conceptual logics suggest that a moral theory can approach one end of this spectrum only by distancing itself from the other. If a theory of morality is understood to involve genuine value and absolute prohibitions, it will stand a better chance of being recognized as a genuine theory of morality. But defining morality in this way rules out scientific demonstration, since value, duty, rights, and the like resist empirical detection. On the other hand, scientific approaches that attempt to reimagine morality in empirical terms stray from an understanding of morality that is adequate to its lived experience, to the genuinely prescriptive character of moral life. The consequence is that scientific approaches at the empirical end of the spectrum fail to persuade people that their conclusions are really about morality. The hope of resolving moral disagreement by appealing to scientific research therefore faces an internal barrier, since moral disagreements appear not to turn on issues that admit empirical resolution.

PART IV
Enduring Quandries

CHAPTER 8

The Quest, Redirected

THE LONG EFFORT to find a scientific foundation for morality has not come to an end. Yet it appears to have reached a certain stasis, framed by the conceptual architecture of the three main schools of Enlightenment thinking on this matter: the psychologized sentimentalism of Hume, the evolutionary account of the mind from Darwin, and the utilitarian, instrumentalist approach of Bentham and Mill. To be sure, the synthesis of these elements is novel, as are the technologies that aid its pursuit, but in terms of the conceptual tools that guide contemporary efforts to fix a scientific foundation for morality, the apparatus is by now conventional. It is within this paradigm that moral psychology and neuroscience operate and where experimental insight accumulates.

Yet there is more to the story. Though the philosophical and conceptual scaffolding is commonly accepted, two important turns have taken place in the footings on which that scaffolding is erected—in the presuppositions of the entire enterprise of finding a scientific foundation for morality. These turns are widely shared, even if not widely acknowledged, and their significance is far-reaching. They bear on what moral scientists now take to be the ontological nature of morality itself.

The first turn is a narrowing of the concept of morality in order to comport with an increasingly unquestionable philosophical naturalism. The second turn is partly a consequence of the first: the new definitions of morality leave out what is both historically and intuitively most essential to morality; in effect—even if not by intent—it leads the new moral science to moral nihilism.

THE FIRST TURN:
TOWARD A DISENCHANTED NATURALISM

The first turn is a radicalization in the naturalism that underwrites the pursuit. Historically, philosophical naturalism has come to suggest certain foundational theses about morality.

First, *the concepts and categories of ethics and morality come from us.* Right and wrong, good and evil, and so forth are human constructs that derive from human evolutionary history, the cognitive architecture of human language, neurochemistry and neuroanatomy, and contingent human interests. Thus the fundamental source of morality is not outside of human experience and biology. There are no real rights, duties, or valuable things out in the world. In the end, Owen Flanagan writes, "We are biological beings, living in a material world that we have constructed. Our norms and values are designed to serve our purposes as social mammals living in different social worlds."[1] Philip Kitcher put it this way:

> We can liberate ourselves from mysteries about many of our current practices by emulating Darwin: think of them, too, as historical products. The aim of [my argument] is to pursue this program in the case of ethics. Ethics emerges as a human phenomenon, permanently unfinished. We, collectively, made it up, and

have developed, refined, and distorted it, generation by generation.[2]

Clearing away the accretions of history and culture, what we find is that "moral rules are born in human minds."[3]

Second, *the fact that moral attitudes feel and seem binding or authoritative can be explained psychologically and neurologically.* The nature and quality of moral attitudes—thinking, feeling, or believing that something or other is either moral or immoral—can be explained psychologically and culturally. Philosopher and legal theorist Brian Leiter, while not one of the new moral scientists, is a naturalist of this stripe, and he helpfully compares moral dispositions to matters of taste—for instance, the preference for Japanese over Thai food. Moral attitudes, he says, "are more ambitious in scope . . . but they are not, on the naturalistic view, different from the gustatory attitudes in their metaphysical or epistemological status."[4]

Yet we experience the moral differently. Accounting for this difference is a matter of speculation,[5] but in the end it can be explained by neural circuitry. As Churchland writes,

> In all animals, neural circuitry grounds self-caring and well-being. These are values in the most elemental sense. . . . In brief, the idea is that attachment, underwritten by the painfulness of separation and the pleasure of company, and managed by intricate neural circuitry and neurochemicals, is the neural platform for morality.[6]

Michael Gazzaniga echoes the point: "moral ideas are generated by our own interpreter, by our brains, yet we form a theory about their absolute 'rightness.'"[7]

Third, *moral intuitions are untrustworthy*. Another way to say this is that moral claims need independent justification. They have no legitimacy on their own terms. For the new moral scientists, moral judgments are best made when based on empirical—scientific—evidence rather than feelings or intuitions. Or at least, they should be based on secular reasoning or calculation.[8] The reason, as Churchland put it, is obvious: "Intuitions . . . are products of the brain—they are not miraculous channels to the Truth."[9]

Paul Thagard is a bit less revisionary, permitting a role for moral intuitions though taking them to be a species of emotional brain process. But even here, moral intuitions are permitted only if a place can be found for them within the naturalistic metaphysic.[10]

Enchanted versus Disenchanted

As we have seen over and over, the view that everything arises from and can be explained by natural properties or causes is by no means novel. Historically, there was room for debate about the meaning of those properties and nature of those causes. That debate continues in the mainstream of science and philosophy today. But, in the new moral science, those properties and causes are routinely understood in exclusively empirical terms. In this discourse, they have been entirely "disenchanted."

Our use of the term "disenchanted naturalism" draws from Max Weber's notion of a thoroughgoing secular, pragmatic, means-end rationality through which, as he put it, the "most sublime values have retreated from public life."[11] In a disenchanted universe, everything can be explained either in Newtonian terms—by lawful, mathematically representable relationships between empirically detectable properties—or, better yet, in mechanical terms, by way of the operations of simple machinelike systems and aggregates of systems. Paul Thagard puts the matter clearly:

> We should judge reality to consist of those things and processes identified by well-established fields of science using theories backed by evidence drawn from systematic observations and experiments. This view ... rules out both religious faith and a priori arguments as sources of knowledge about reality. . . . Reality is what science can discover.[12]

By contrast, reality is *enchanted* if it also includes nonphysical, nonempirical phenomena not well-suited to quantitative representation or description by laws of nature. The trouble with an enchanted universe is that it introduces elements that can't be captured in either Newtonian or mechanical terms. This is regarded as a problem because things that aren't explicable in these terms may seem to be "spooky," "occult," or "mysterious."[13] The thought is that in failing to be Newtonian or mechanistic, phenomena become difficult to understand, much as ghosts and other "spooky" things presumably would be if they existed.

We want to be clear here: enchantment doesn't require commitment to deities, spirits, or anything *super*natural. Belief in supernatural phenomena is certainly sufficient for an enchanted view of reality, but is by no means required. It is unexceptional within mainstream metaphysics and philosophy of mind that one can posit naturalistically unacceptable entities—souls, abstract universals, libertarian free will, and so on—without accepting anything supernatural.

With this clarification in mind, the dispute between enchanted and disenchanted naturalism occurs along many fronts. Most of these disputes turn on whether a certain phenomenon, if taken to be genuine, would be too spooky or mysterious to be naturalistically acceptable. To convey a better sense of what sorts of phenomena might be considered enchanted—and to better illustrate the nature of the conflict between enchantment and disenchantment—here are a few examples:

- **Life.** Is life really a sui generis active principle, or is it an illusion brought about by aggregates of molecular machines, working in concert, shaped by natural selection?[14] The enchanters incline toward taking life as a real, active thing unto itself while the disenchanters are inclined to say that it is reducible to chemical processes.
- **Intentionality.** Are any words, pictures, or thoughts genuinely *about* that which they appear to represent or refer to? Or is the apparent "aboutness" of thoughts and words a mere correlation? In other words, is intentionality real? The enchanted naturalist says yes, the disenchanted naturalist, no. The disenchanters say no in part because it is hard to see how to make sense of one thing really being *about* another if you can only appeal to the basic properties of physics to build what there is—spin, mass, charge, quark color, and so on.[15]
- **Free will.** Can we ever really *decide* between options, or are we always determined to do as we do by impersonal causal forces outside our control and of which we are ignorant? That is, are we ultimately *agents* or are we *machines*? Here again, the enchanters might say people have a genuine agency while the disenchanters would say that we have little to none.
- **The self.** Is each of us a self? Is there really some unified thing that has your subjective experiences? Those on the

enchanted side say yes. Those for disenchantment may say no. Following roughly in Hume's skeptical tradition, the disenchanters say that the self is an illusion and that what we think of as our self is instead just a collection of many discrete cognitive processes.[16]

▸ **Consciousness.** Our subjective experience of what things are like seems not to be explainable in terms of the objective features of our brains. Given the apparent lack of a natural explanation, is our conscious experience real? The disenchanters would say no—conscious experience is an illusion;[17] the enchanters say yes.

▸ **Purposiveness.** Does anything have a genuine purpose or *telos* beyond the purposes human beings have for them? Is there any teleology not born from human intentions? The enchanters might say yes, the disenchanters, no.[18]

What about morality or value? Are there really things we *should* and *shouldn't* do beyond what would best serve our interests and preferences? Does reality include the sort of normative force that moral agents are capable of feeling? Are some things valuable in an objective sense, beyond what we happen to want or care about? The enchanters say yes. But these phenomena would be spooky indeed, and the disenchanters say no.

Within a disenchanted naturalism, there can be no irreducible moral "oughts"; there is no fundamentally *moral* normativity. As Leiter put it, "Real normativity does not exist: *that is the entire upshot of the naturalist view*."[19] Similarly, Greene argues that nothing in reality could explain the truth of moral claims.[20] What had long been a suspicion in modern philosophy has now become a creed: morality isn't real.

Moral scientists who depart from this disenchanted view are rare, the most prominent being Sam Harris. While in other ways a committed naturalist, Harris nevertheless affirms an irreducible moral reality.[21] But he is atypical.[22] As a general rule, the new moral scientists do not accept that genuine morality is any part of reality.

That said, they recognize the notion that there are things we ought and ought not to do, but what they think "ought" means is the same thing that is meant by other, *nonmoral* "oughts." Owen Flanagan describes the position like this: "Ethics is not a separate domain of knowledge—it's probably continuous with prudential and practical knowledge. . . . It just has to do with matters . . . we take to be of greater importance."[23]

So, for example, if you want to collect every issue of a given series of comic books, you ought to buy the one issue you've yet to obtain when available. Similarly, there is a sense in which you ought not rest your feet on the table in a restaurant. In the first case, the ought is a practical or prudential one—it concerns what you should do to satisfy your preferences. The second ought relates to rules of etiquette.

But neither of these oughts captures what we typically mean by the term in everyday moral contexts. Most people, for example, believe that one ought not murder people. That sense of "ought" goes far beyond etiquette or prudence; even if you knew you could get away with murder, it is still something you ought not do.

The disenchanted naturalist, however, doesn't think such a distinction can be maintained. For the disenchanted naturalist, there are only the other kinds of "ought." If we want to continue to classify certain things as actions that "ought not be done," we must understand these oughts in terms of preference or prudence.[24] The prevalence and plausibility of this view makes sense when seen in light of the enduring significance of G. E. Moore's contention (from his naturalistic fallacy) that goodness cannot be a natural property. Combined with the view that everything that exists is natural, Moore's contention forces the conclusion that goodness doesn't exist.[25]

The new moral science does recognize that human beings have moral *feelings*: it feels to us as though certain things are good and bad, right and wrong, appropriate and inappropriate, and so on.[26] Moral language develops because it is useful both for describing our moral feelings and for solving problems. Even if there isn't *really* any morality, we find some actions better or worse than others, and find moral language a useful way to talk and reason about these beliefs. But this language comes from us. While some of its components were created intentionally, most of it, in all likelihood, arose by convention during our evolutionary and cultural development.

Here a clarification is in order. *We are not saying that practicing scientists accept this sort of disenchanted naturalism in any universal way.* While some surely do, our target isn't scientists as such, but instead is the new moral scientists and their surrounding discourse. And in this context, the prevailing view is—as we show through extensive citation and explanation of basic perspectives—that everything must reduce to scientifically describable entities and forces.

There are also those naturalists who hold that these enchanted phenomena—mind, morals, life, and so on—*emerge from* but are not *reduced to* or *eliminated in favor of* basic physical phenomena. However, as we show, the main protagonists of the discourse do not accept this. And for what it is worth, no one in fact has any idea *how* enchanted features emerge from scientifically tractable reality. It's a fundamentally hopeful posture, but given the apparent chasm between the realms of physical reality and that of experience, it is as yet unclear whether the idea is workable.

The Second Turn:
The Original Quest Abandoned

The second turn, which follows from the first, amounts to a crucial departure from the original quest to establish an actual scientific foundation for morality.

Think about it historically. As we have argued, overcoming moral disagreement is one of the chief motivations for the pursuit of a scientific morality; an imperative made all the more pressing by the violence and harm that moral disagreement has often brought about. But another critical part of this effort historically was the concern that moral disagreement might devolve into moral skepticism and an eventual abandonment of moral reality altogether. Moral theorists in the seventeenth century worried that moral disagreement would lead people to conclude that moral reality was finally unknowable, or at least indemonstrable. Grotius's appeal to evident and objective bases for moral law aimed to stem the tide of moral skepticism. In the same way, Locke's empiricist epistemology was intended, in part, to lay the groundwork for a moral philosophy that could persuade moral skeptics.[27] Thus a founding objective of the science of morality was to secure a empirically obvious basis for morality that would both resolve disagreement *and* restore belief in objective moral truth.

Today, however, few of the new moral scientists think that science can stave off moral skepticism. Ethical and moral properties—conventionally understood to include irreducible ethical value, basic rights and duties, and uniquely prescriptive normative force—have not been connected in the right way to what the sciences study. So they have ruled these ethical phenomena out of existence. With rare exception, the new moral scientists have abandoned this central part of the quest; most now claim that moral reality simply does not exist, except as an arbitrary construct of social, psychological, and biological life.

As we have seen, Hobbes and Hume foreshadowed this conclusion. Rather than focusing on the nature of binding, prescriptive moral laws, Hume attended to what humans do when making judgments about moral questions. His theory aimed to identify lawlike relationships between human behavior and moral judgments—in particular, the human tendency to identify certain character traits as virtuous or vicious. Hume's moral psychology fit uneasily with any view on which morality is a genuine and objective feature of reality, independent of human judgment. Humean views permit us to see morality as just another element in the realm of contingent human perspective.

But in his time, Hume's rethinking of the nature of ethics didn't stick. Most found his argument simply beyond the pale. As Sidgwick observed in 1886,

> The fundamental questions "What is right?" and "Why?" tended to drop somewhat into the background—not without manifest danger to morality.... A reaction, in some form or another, against the tendency to dissolve ethics into psychology was inevitable.[28]

To Sidgwick, the reaction to this "manifest danger" went in two directions: one, in the development of utilitarianism by Bentham and Mill; the other, in a recommitment to an intuitive basis for moral knowledge. Importantly, both of these reactions deliberately moved away from Hume's psychologistic account and toward the affirmation of moral reality—in the case of utilitarianism, the affirmation that "conduciveness to pleasure is the ultimate standard of morality"; in the latter, the affirmation of the "objective validity" of "ultimate ethical truths."[29] Thus, while Hume's ethics influenced the work of his contemporary Adam Smith, by and large, Humean theory receded from mainstream intellectual discourse on the subject. Not until the late nineteenth and early twentieth centuries, aided first by Darwin's natural history of the human mind and later by logical positivism's thoroughgoing naturalistic approach to ethics, did it regain credibility.

What few intellectuals would once embrace publicly has now become de rigueur for the new moral science. Today, the Humean psychological view of morality no longer seems radical or extreme. It has become so commonplace that we may easily miss the significance of the change.

From Moral Skepticism to Moral Nihilism

Historically, moral skepticism took shape as the belief that moral truth cannot be known. With the new moral science, however, the moral realm is now something exclusively demarcated by our neurochemical impulses and the legacy of our contingent evolutionary development. There is no room for the genuinely prescriptive, for real value or obligation. *This goes beyond the belief that moral truth cannot be known, to the belief that there is no genuine morality. If the former is moral skepticism, the latter is moral nihilism.*

It is important to note that most practitioners of the new moral science do not call themselves "moral nihilists." But intentions are mostly irrelevant if, in fact, that's what the view is.

While many practitioners of the new moral science share the underlying metaphysics of a disenchanted naturalism, most do not explicitly state how damaging this metaphysics is to an ordinary conception of morality. Among the new moral scientists, Joshua Greene has been explicit about the moral implications of his naturalistic metaphysics, beginning with his dissertation at Princeton.[30] There he argued for "metaphysical skepticism" about morality, the idea that reality cannot possibly include genuine moral phenomena:

> Moral realism requires the existence of true moral principles, claims that ascribe moral properties to things in virtue of their value neutral properties. . . . We eliminated each of the possible types of fundamental, necessarily true moral principles, leaving us with the surprising conclusion that there are no true moral principles and that moral realism is false.[31]

Greene writes off moral realism not through argument or evidence but because he personally "cannot take such views seriously" and has "little to offer in the way of convincing arguments to those whose inclinations are otherwise." He recognizes that some might accuse him of rejecting nonnaturalistic ethics simply because those views don't fit his naturalist assumptions, but he denies this, saying that his "commitment to naturalism has . . . amounted to little more than a refusal to take seriously those views that invoke God, Moorean non-natural properties, Platonic forms, and the like."[32] Yet the common factor in the ethical views he rejects is that they are nonnaturalistic. As such, he appears to reject them precisely because they do not fit with his naturalist assumptions. A decade later, in *Moral Tribes*, Greene softened his stance, arguing in the footnotes that while he retains his view that there is no objective right or wrong, this issue isn't what matters most for practical purposes.[33]

Alex Rosenberg has been even more outspoken about what his view implies for genuine morality. Rosenberg, the R. Taylor

Cole Chair in Philosophy at Duke, says, "In a world where physics fixes all the facts, it's hard to see how there could be room for moral facts. . . . Why bother to be good? . . . We need to face the fact that nihilism is true."[34] For Rosenberg, moral nihilism is the logical conclusion to the view that the world is reducible to physics. Empirically based science rules out genuine ethical value and duty. But there is more to his claim. His argument begins with two premises:

> First premise: All cultures, and almost everyone in them, endorse most of the same core moral principles as binding on everyone. Second premise: The core moral principles have significant consequences for humans' biological fitness—for our survival and reproduction.[35]

Rosenberg then points out that the confluence of "core morality" and reproductive fitness could not have happened by chance:

> A million years or more of natural selection ends up giving us all roughly the same core morality, and it's just an accident that it gave us the right one, too? Can't be. That's too much of a coincidence.

But if the confluence of our core morality and reproductive fitness isn't a coincidence, then what makes core morality the right and true morality must be that it leads to reproductive fitness. But that, he insists, can't be right either.

> Which is it? It can't be either one. The only way out of the puzzle is nihilism. Our core morality isn't true, right, correct, and neither is any other. Nature just

seduced us into thinking it's right. It did that because
that made core morality work better; our believing in its
truth increases our individual genetic fitness.[36]

This argument from coincidence is unique to Rosenberg, at
least among the new moral scientists we focus on. But, again, the
cornerstone of his argument is a disenchanted philosophical naturalism. Because he takes there to be ultimately nothing besides
matter in motion, he concludes that evolutionary selection could
not have guided the development of core morality.[37] In the end,
Rosenberg insists that "We have to acknowledge (to ourselves,
at least) that many questions we want the "right" answers to just
don't have any. . . . Alas, it will turn out that all anyone can really
find are the answers that they like."[38]

Though he is less explicit about it than Rosenberg and Greene,
this is a clear implication of Philip Kitcher's position as well. He
claims that we shouldn't accept the existence of anything that
science can't, *in principle*, discover. As he put it,

> Fundamental to naturalism is the desire not to multiply
> mysteries—not beyond necessity, but not *at all*. . . . This
> does not entail that the inventory of things they accept
> is restricted to the list of entities countenanced by cur-
> rent science.

Rather, he allows that since current science is incomplete, the
naturalist must make room for future discoveries. Philosophers
should be allowed to postulate new kinds of things, Kitcher
asserts, but the standards for what should be taken seriously
should be

the standards our most rigorous investigations set for themselves. *There are to be no spooks.* That means no invocations of the Forms, or of Natural Law, or of Processes of Pure Practical Reason, or even of Moral Properties accessible to ordinary human faculties—*all have to be shown to accord with standards of reliable inquiry, or they have to go.*[39] (emphasis added)

Since Kitcher doesn't think moral entities can be shown to accord with the standards of "our most rigorous investigations," one can only conclude that they don't exist.

Owen Flanagan, too, operates with a disenchanted naturalistic account of human life, on which "whatever we are, or turn out to be, cannot depend on possessing any capacities that are not natural for fully embodied beings."[40] Flanagan claims that "the very idea of the human sciences implies that all human practices can, in principle, be understood scientifically"[41] and, in discussion of this idea, adds that "science can, in principle, explain everything we think, say, and do—that it can, in principle, provide a causal account of human being."[42] Yet Flanagan claims to be agnostic about moral reality:

My kind of ethical naturalism implies no position on the question of whether there really are, or are not, moral properties in the universe in the sense debated by moral realists, anti-realists, and quasi-realists.[43]

It raises the question, how one can be committed to a view that one has no explicit opinion about. Yet it can happen: suppose you had never thought about whether there was a God, but you believe that there are no supernatural beings of any kind. Your position commits you to there being no God, even though you don't have an explicit view about that. Minimally, this appears to be Flanagan's position with respect to genuine morality.

Patricia Churchland also appears to embrace this metaphysical skepticism about genuinely prescriptive morality. While she is difficult to pin down explicitly, admitting that her naturalism rules out this sort of morality, she says things that imply this. For instance, "It is increasingly evident that moral standards, practices, and policies reside in our neurobiology," and

> the traditional field of ethics must itself undergo recalibration. Philosophers and others are now struggling to understand the significance of seeing morality not as a product of supernatural processes, "pure reason" or so-called "natural law", but of brains—how they are configured, how they change through experience, how cultural institutions can embody moral wisdom or lack of same, and how emotions, drives, and hormones play a role in decision-making.[44]

Recall also from chapter 7 her assimilation of the moral to the social—understanding morality socially and prudentially, but not as a source of objective action-guidance.

Stepping back a bit from these particular theorists, if there is no morality, if nothing is genuinely valuable, if there really is nothing beyond human preference, convention, or etiquette to frame our decision-making, then it would seem that any course of action is, intrinsically, as good or bad as any other.[45] Even if one can provide a sociological or psychological explanation for it, any practical proposal would, *in fact*, be morally arbitrary.

Redefining Morality

Though the new moral science has abandoned the quest as originally understood, it has not lost the ambition to establish something that resembles a science of morality. To do so, however, the moral scientists have to change the terms of their project as well as their tactics. They have to redefine and redirect what it means to have a science of morality. Greene lays out the challenge elegantly:

> Perhaps our moral questions have no objectively correct answers. But even if that's true, knowing that it's true is not much help. Our laws have to say *something*. We have to choose, and unless we're content to flip coins, or allow that might makes right, we must choose for *reasons*. We appeal to some moral standard or other.[46]

In effect, Greene would say, "Pay no attention to the yawning abyss at the foundations of my preferred moral theory—focusing on that won't help us." Perhaps not. But keeping that abyss in mind does help clarify what Greene and the other new moral scientists really mean when they use moral language and invoke moral concepts in their recommendations for what we should do. Unless one is paying careful attention, it is easy to read them as operating within conventional, prescriptive moral language and moral systems. But they are not.

In fact, Greene and others propose that we should revise what moral language means. Fiery Cushman writes that "by unmasking our minds as the authors of morality, we may be better able to bend its narrative arc toward a happy end . . . [and] gain a magical power of control over its future."[47] Greene argues that we should move away from thinking of moral concepts in terms of right and wrong, and instead see them as serving human interests.[48] This is an important move because it reframes the terms by which morality is understood. Though the language is the same, the meanings of moral terms have been changed into something different: something that superficially resembles common or conventional morality but in fact isn't at all like it.

When Greene urges us to "appeal to some moral standard or other," he can't be referring to a *genuine* moral standard because he has already denied its existence. All that remains to ground reasons for human action are the ultimately nonmoral arenas of what are sometimes called "moral" prudence (how to get what we want) or social convention (what is typically done). When they render morality into something more amenable to naturalistic explanation, the meaning of morality is thus changed into something different—into mere prudential or conventional concerns.

As we saw in chapter 4, this is the direction Owen Flanagan takes. "We are looking for norms, values, and practices that are the best, where 'the best' is almost always 'the best for such and such purpose or purposes.'"[49] In effect, he has redefined the realm of morality as the realm of practical reason. He says—as stated before—"Ethics is not a separate domain of knowledge—it's probably continuous with prudential and practical knowledge. . . . It just has to do with matters . . . we take to be of greater importance."[50]

Part of what makes this interesting is that he calls this ethical theory "eudaimonics"—the empirical study of human flourishing.[51] Flanagan's eudaimonic theories bear some structural similarities to Aristotle's ethical view, which was an account of prescriptive morality. Thus it would be easy to come away from reading Flanagan's work thinking that he too accepts that prescriptive morality is a genuine phenomenon. But a closer reading reveals it to be otherwise. Rather than a condition of being, morality is a state of mind; rather than objective and true, it is functional and manipulable; rather than constitutive of an objective human flourishing, it is useful for fluid conceptions of human well-being. Alex Rosenberg comes right to the point: "reducing moral rightness to prudence produces a naturalistic grounding for morality by *changing the subject*."[52]

The new moral scientists are not alone here. Today in philosophical metaethics—the study of the foundations of ethics—the nature of morality is an open question. The new moral scientists merely take one of several possible stances in this contested realm. No one is conspiring to deceive. Yet there is a confusing ambiguity in the terminology underlying their appeal to the public. By talking about morality as if it were what most people assume morality to be, in effect the new moral science perpetrates a bait and switch.

Grounded in the Current Social Consensus

On what basis, then, does this prudential ethics rest? Greene is clear: "We've no choice but to capitalize on the values we share and seek our common currency there."[53] Elsewhere he writes, "We're looking for a [common moral standard] based on shared values. For our purposes, shared values need not be perfectly universal. They just need to be shared widely, shared by members of different tribes whose disagreements we might hope to resolve by appeal to a common moral standard."[54] The best strategy is:

> to seek agreement in shared values. Rather than appeal
> to an independent moral authority . . . we aim instead to
> establish a *common currency* for weighing competing values. This is . . . the genius of utilitarianism, which estab-

lishes a common currency based on *experience*....[55] We can take this kernel of personal value and turn it into a *moral* value by valuing it *impartially*. . . . Finally, we can turn this moral value into a moral *system* by running it through the outcome-optimizing apparatus of the human pre-frontal cortex. This yields a moral philosophy that no one loves but that everyone "gets"—a second moral language that members of all tribes can speak . . . This is the essence of [my view]: to seek common ground not where we think it ought to be, but where it actually is.[56]

In the end, the utilitarianism that Greene and the other new moral scientists advocate isn't driven by the greater value of some outcomes over others—this isn't the utilitarianism of Mill or Sidgwick or any other moral realist. Rather, it is driven by what people *happen to value* at the time (and maximizing this), rather than *what is intrinsically valuable*.

Flanagan echoes this view, giving us the metaphor of a bridge across a river. Where did this goal or end come from?

The answer is this: *It came from the people*. Enough of them shared the goal of doing business with the folks on the other side of the river that seeking to meet this goal was judged to be a good idea. The same is true, I claim, for what seem to be, but are not, the more mysterious kind of normative questions that philosophers fuss with and often mystify.[57]

Here, too, the basis for deciding whether we should do something is whether doing it contributes to a goal that enough people share.

Framed in this way, the quest for ends or goals is paramount.

But what are our ends or goals? Though the historical record clearly shows how widely they vary, Greene thinks he knows. "We all want to be happy. None of us wants to suffer. And our concern for happiness and suffering lies behind nearly everything else that we value."⁵⁸ Thus "our task, insofar as we're moral, is to make the world as happy as possible, giving equal weight to everyone's happiness."⁵⁹

On Greene's view, happiness is the overall quality of a person's experience. Since many things can affect our experience, he holds that the sources of happiness can be quite varied, including family, friends, love, personally meaningful work, and character.⁶⁰ And as opposed to the thin, crude happiness of Bentham, Greene recognizes that the sorts of things that improve the quality of our experience don't all count the same—push-pin isn't the equal of poetry. As such, his view appears to overlap with thicker, more robust, and fairly prevalent conceptions of happiness.

But—crucially—Greene's view isn't that we should try to make everybody happy because happiness is such a valuable thing— remember, Greene doesn't really think anything is valuable. Instead, his view is that we should try to make everybody happy because that's what *everybody wants*.

This might seem like an inconsequential distinction, but it isn't. What people happen to value and what really is valuable are two very different things. After all, just because people *want* something—even if *everyone* wants it—that doesn't make pursuit of that thing acceptable. Some people value hoarding old newspapers, but does this make hoarding old newspapers really valuable? This is unconvincing. At the same time, it may be that some things—say, for instance, human life—would be valuable even if no one valued it. But in the absence of any real value, Greene merely seeks to maximize what people *happen to value*.⁶¹

So we see that Greene makes happiness the main object of human action not because he thinks happiness really is the greatest good. Greene doesn't think anything is good at all, literally speaking. Why then prefer happiness? His overall project is to propose a solution to moral conflict between groups with different moral views. His strategy is twofold: first, to reject moral pluralism as it exists on the ground by ignoring the particular claims of competing groups in the particular moral languages in which those claims are made; and second, to find something everybody values and build all of ethics on that common basis. He believes that happy experiences are this common basis, and hence his choice of happiness-based utilitarianism.[62]

Greene is hardly alone. Happiness has become the rage in popular culture not least because it is legitimized by the burgeoning field of positive psychology. Psychology has hardly abandoned pathology, but it has also turned to the project of discovering what makes people happy. Here, too, morality is redefined from historical categories into the categories of utility, functionality, and capacity as they bear on subjective well-being.[63]

Rendered as self-reported happiness, morality becomes amenable to a certain kind of empirical analysis. As Flanagan notes, this doesn't make the study of morality scientific "in the modern sense . . . ," though it does involve "systematic philosophical theorizing that is continuous with science and which therefore takes the picture of persons that science engenders seriously."[64]

FROM DISCOVERY TO INVENTION

The modern natural lawyers, people like Grotius and Pufendorf, were looking to discover principles of morality that would persuade others that particular claims were the correct, justifiable moral principles, binding for all humanity. Their model was the hard sciences, where demonstrations were able to resolve disagreements about which claims about the natural world were objectively true. The hope was for a verifiable *discovery* of morality that would similarly end debate. But now, centuries later, practically all proponents of the new moral science have abandoned that quest. Locating and demonstrating a realm of justifiable moral principles has been ruled out as impossible.

By the lights of disenchanted naturalism, in which there is no such thing as a moral *should*, the question of what we should do becomes nonsensical. Any recommendation of action or policy *must be* morally arbitrary. There can be no moral basis for preferring any course of action over any other. Some actions will still be better or worse with respect to what someone is trying to accomplish, or with respect to adherence to conventional rules, but *better* and *worse* here carry no moral weight. These terms have no normative force: plans can be better or worse given the goal of robbing a bank. *Better* and *worse*, understood in this way, are good or bad only insofar as the goal in question is good or bad. But good and bad goals—as real, objective states of affairs—are exactly the sorts of things that don't exist inside the disenchanted naturalism of the new moral science.

Since the existence of any real ethical basis in reality has been denied, there is no ethically relevant rational basis on which Greene and other practitioners of the new moral science can argue that their view is in any sense objectively better than that of the ISIS caliphate or a Russian oligarch. It seems their position must, in the end, come down to mere preference, or the preferences that a perceived social consensus would allow. In this light, maximizing happiness is just as good a choice for what humans should strive for as anything else that social consensus would allow. It could just as easily be something else.

Taking the Measure of Our Dilemma

The quest for a demonstrable scientific foundation upon which to base the moral and social order has fundamentally changed in the late twentieth and early twenty-first centuries. The new moral scientists no longer seek scientific discovery of moral truth, for under the metaphysics of disenchanted naturalism, moral truth does not and cannot exist. Rather, they argue that we can use science to help us achieve our societal goals, recognizing that these goals are, at best, arbitrary.

The logical consequences of this are difficult to explicitly own. Rosenberg and Greene are the exceptions. But whether any particular scholar acknowledges it or not, it is hard to avoid the conclusion that the cultural logic of the new moral science inevitably leads to moral nihilism.

And most of us are loathe to admit it ourselves. Against the pervasive self-interest of our market culture, the inescapable cynicism of our political culture, and the extensive vacuity of our entertainment culture, we want to believe that there is *some* foundation upon which we as a nation or world can make strong moral claims; claims that will inspire us, guide us, and unify us. Perhaps this is why so many overlook the blurring of lines between "is" and "ought" or minimize the tendency by some moral scientists to overreach by making grand moral claims upon modest if not dubious descriptive findings. To look to religion for such a unified, socially binding foundation is, of course, out of the question. After our long journey through the centuries, we cannot escape the conclusion that science fails as well.

CHAPTER 9

The Promethean Temptation

AND THE PROBLEM OF UNINTENDED CONSEQUENCES

O UR STORY of the quest for a science of morality has come to an end. Condensing the narrative down into a couple of pages, what have we found?

A SUMMARY OF THE NARRATIVE THUS FAR

Grotius, Hobbes, and other early modern thinkers took the first steps toward a scientific account of morality. Inspired by the Scientific Revolution and its metaphysics, and zealous for a moral standard other than those claimed by warring religious factions, Grotius's rights-centered approach to ethics would prove influential but insufficiently empirical to found a science of morality. Hobbes's mechanistic psychology of moral thought, though speculative rather than empirical, paved the way for similarly naturalistic approaches to come. Shortly thereafter, Newton's simple, lawful explanation of various physical phenomena became the gold standard for explanation—an ideal widely sought after in the Enlightenment.

This is especially true of Hume and Bentham, who both sought scientific moral theories, though differing widely in their approaches. Just as Newton had explained celestial mechanics with simple laws connecting the basic features, so did Hume attempt to give simple, lawful explanations for the interactions between human traits and human moral sentiments of approval or dissent. Bentham's utilitarianism emphasized the role of calculation in determining which actions most contribute to positive outcomes, rigorously defined, and he left no room for traditional but less empirical aspects of morality, such as duties and goods not based in pleasure.

Though Hume was largely forgotten until the twentieth century, Enlightenment scientific accounts of morality persisted into the nineteenth century through Bentham and refined by Mill. Darwin's idea of natural selection provided the basis for a satisfying explanation for the origin of human moral sentiments, and for moral psychology generally—but this wouldn't become clear until the middle of twentieth century. The immediate ethical applications of Darwinism were fallacious—Spencer's "evolutionary ethics"—and were corrected in the late nineteenth and early twentieth centuries by Moore and others.

In the early twentieth century, increasing disciplinary isolation, along with behaviorist trends in psychology that ignored evolution, prevented a coherent scientific picture of human morality from forming until the late-twentieth century. But breakthroughs in evolutionary biology, the cognitive revolution in psychology, the fearless though naïve philosophizing of E. O. Wilson, and new technologies for studying the brain encouraged new focus on a comprehensive empirical understanding of human morality. These elements have been refined and supplemented by philosophers, incorporating accounts of ethics that best dovetail with the prevailing empirical picture of the moral mind. Hume's mind-focused sentimentalism, Darwin's evolutionary account of the mind, and an instrumentalist ethical methodology bearing strong affinities to the utilitarian calculus of Bentham and Mill are now all embedded within a disenchanted naturalism that is committed to empirical study of the mind. This new moral synthesis has been a significant development in recent academic history.

But it has failed to deliver on the longstanding hope for a scientific foundation for morality. The interdisciplinary coherence of the new synthesis picture of morality has not brought us any closer to a consensus-making scientific account of morality. Instead, it has delivered increasingly detailed descriptions of the physical apparatus involved in human moral thought and action. These descriptions range from explanations for biological altruism in evolutionary biology to plausible hypotheses concerning the evolutionary development of human empathy in primatology and evolutionary psychology.

These are valuable additions to our knowledge of evolutionary biology and psychology. But as descriptive analogues of prescriptive moral concepts, these empirical successes do little to advance our knowledge of morality. Not that the scientists themselves are always responsible for the sense that their discoveries answer deep moral questions. Some blame goes to the overzealous and sensationalistic packaging by media reports of sober scientific work. But the practitioners too sometimes overreach. We described some of the more notable and reckless attempts to leverage modest scientific findings into substantive moral conclusions. Even when moral scientists stop short of overreaching, they often rely on blurred boundaries to exploit the interest in a substantive moral science and yet, when pressed on the point, excuse their work as merely descriptive even though it is couched in the language of morality.

Can the science of morality deliver on its original promise? There are those who say it eventually, inevitably will.[1] The problem is that this kind of "promissory naturalism" imputes a teleology to science that science cannot justify. And there are formidable challenges to getting there.

There is the challenge of definition: of providing a clear and concise definition of morality that is adequate to lived experience. And there is the challenge of demonstration: of construing morality in ways that are scientifically measurable. As we have seen, these operate at cross-purposes: the more of one, the less of the other. The more morality is defined in an empirically measurable way, the less it resembles morality as it seems to be, and the more faithful the definition of morality to experience, the less it is empirically measurable.

These challenges are formidable, possibly insurmountable. While it is safe to conclude that the quest has produced a modest science of aspects of the physical apparatus underlying human morality, it has not—even on its own terms—provided anything like an actual science of morality. The fragments sometimes held forth as a science of morality are constrained by conceptual flaws, evidential overreach, faulty logic, and arbitrary assumptions.

Yet a closer examination of the recent contributions to this science, especially those growing out of the new synthesis view of morality, reveals that a significant change has taken place. It seems fair to say that much of the current science of morality is no longer really a science of *morality*. Instead, part and parcel with the metaphysics of philosophical naturalism, the bulk of the new moral scientists do not think there is any such phenomenon as morality, as traditionally conceived. A kind of moral nihilism has taken hold, accepting no genuine, justifiably action-guiding *right* or *wrong* or *good* or *evil*. There is nothing anyone really *should* or *ought* to do, strictly speaking. Such phenomena, if they exist, can't be detected or studied by the harder sciences, and hence we lack reason to think they exist. This logic of disenchantment threatens much more than morality. It separates the scientifically pure concepts from unclean ones such as consciousness, intentionality, life, free will, and the like. Questions of action, of policy, of normative guidance are thus no longer moral or ethical matters.

What is confusing is that the language of morality is preserved. The new moral science still speaks of what we "should" or "should not" do—but the meaning has been changed. Normative guidance is now about achieving certain practical ends. Given that we want this or that, what should we do in order to obtain it?

The quest, then, has been fundamentally redirected. The science of morality is no longer about discovering how we ought to live—though it is still often presented as such. Rather, it is now concerned with exploiting scientific and technological know-how in order to achieve practical goals grounded in whatever social consensus we can justify. The science of morality, then, has evolved into an engineering project rooted in morally arbitrary goals.

Because genuine good, bad, right, and wrong have been jettisoned, there is finally no basis on which to adjudicate, in any principled way, which ends should be pursued. Of course, no one is saying we are free to adopt any goals or objectives we please. And in their various policy or behavioral proposals, the new moral scientists never fail to recommend those sanctioned by safe, liberal, humanitarian values. With concern, however, we point out that from their own perspective, these values can boast of little more than the historically contingent support of present elites in Western society. By their own commitments, there can be no deeper good that anchors this consensus over any other at any other place or time in world history.

The search for a science of morality, then, is a story of a hopeful quest spanning centuries, many promising ideas, many sobering failures, cycles of dormancy and revival, overreach, and finally, a momentous but unnoticed reorientation away from the original goal and toward a radically different end.

A Failed Quest . . .
Animated by Good Intentions

This is our story. But what does it mean?

We have been careful to avoid the labels "positivism" or "scientism" in this book. They can become tropes that are less descriptive than reductive—labels invoked to dismiss the strong claims of the new moral science without doing the hard work of puzzling through those claims.

Dismissing the claims of the new moral science outright, as many do, has the unfortunate effect of obscuring the longings that animated the quest in the first place and continues to do so today: the Promethean longing for a clarity that rises above confusion, and for sureness in a world clouded with ambiguity and doubt. Even more significantly, it diminishes the longing for a flourishing that takes us beyond conflict, violence, and misery that so often results.

Our own tragic sensibility leads us to the view that as utopian as these ends may be, this is a longing we should nevertheless embrace. The sources of moral conflict are as present as ever, the intention to do harm is as ruthless as ever, the means to do violence are more numerous and powerful than they have ever been, and thus the misery we humans are capable of inflicting on each other and on the earth is greater than ever before. It strikes us as minimally sensible, if not urgent, to search for any humane way through our deepest differences.

But will science get us there?

The story of the quest for a scientific foundation for morality persuades us that the answer is no. At least, science has not gotten us there yet, there are no promising signs that it might, and no plausible solutions to the challenges this project faces.

Science has taught us many things. Applied to the problems of human existence, it has brought about immeasurable benefits in health, longevity, comfort, ease of living, and security. A central part of its achievement is the immense practicality of its method and findings. It urges us to credit, and to build upon, only what can be demonstrated for all to see.

Yet for all that science has taught us and for all the good that it has brought about, it has clearly not provided anything like a solution to the problem of morality—no way of resolving moral disagreement with empirical methods.

The Failed Quest and Its (Mostly Unintended) Consequences

Up to this point, our account, challenges, and criticisms of the science of morality have all been internal to that discourse. That is, they have concerned aspects of arguments or theories well known (or easily accessible) to the moral scientists themselves. We think it is essential, however, to also highlight another set of challenges, which are rarely addressed in the discourse itself.

In chapter 1, we spoke of the dangers of this field's insularity. The truth is that insularity is an occupational hazard within all academic fields and subfields. But for the science of morality, resisting insularity is all the more important in light of its real-world implications.

For all of the erudition and intellectual sophistication to be found in this particular corner of the world of science and philosophy, we wonder whether its practitioners operate with an appropriate self-consciousness about the context in which their work and ambitions unfold. Most of them understand themselves to be disinterestedly pursuing truth. They imagine that they are working from the purity of their own metaphysics, uncontaminated by anything outside of science and reason. We have no wish to impugn anyone's motives; indeed, we take them at their word: they *are* pursuing truth. But they are doing so within the frameworks (and limitations) of knowledge prescribed by the methods and concepts current within their fields of inquiry. And that is not all.

There is a temptation to assume that the world of science or the world of philosophy is *the* world, and that the pursuit of truth somehow rises above the push and pull of social life and the swirling dynamics of the historical moment. What is arguably missing within this discourse is the caution that comes from reflecting on its human and thus contingent nature. One rarely finds in the philosophical or scientific literature any self-consciousness about the ways in which science and philosophy are embedded within a social and historical context and thus operate alongside a range of exogenous—that is, nonphilosophical and nonscientific—social dynamics.

It should go without saying, but apparently it doesn't, that science and philosophy are themselves human products—no less products of history and culture than are politics, ideology, ethics, and theology. Social dynamics surround and permeate these systems of knowledge. Linguistic conventions are simply the most basic of these frameworks, and they are inescapable. Also inescapable are the complex dynamics of power, interest, privilege, ritual, ambition, resentment, injury, and hope—among many other things. Some may be tempted to argue that philosophy and science are privileged epistemologies that are somehow working above the fray. But that argument is, if not absurd on the face of it, certainly debatable. Although they aspire to an ideal of objectivity, the history of science offers a practically endless catalogue of researchers seeing and understanding what they were socially conditioned to see and understand.

A Moral Science Unreflective about History, Culture, and Political Economy

One challenge to the new moral science is the effective absence of any working awareness of or engagement with history, culture, or political economy. Take as an example the centrality of "social consensus" in the new utilitarian logic. Joshua Greene and others speak about the necessity of relying upon "the shared values" of social consensus to build an ethics; that ethics begins with what "the people" want. Surely we should be cautious about that idea. After all, social consensus in 1930s Germany gave us a democratically elected Nazi Party, the Third Reich and its war machine, and in turn, the horrors of the Holocaust. Social consensus gave us a democratically validated slavery, Jim Crow segregation, the terror of lynching, and every conceivable expression of social denigration and discrimination. Social consensus gave us apartheid in South Africa, ethnic cleansing in the Balkans, and genocide in Armenia, Darfur, Burma, Rwanda, Cambodia, Somalia, and the Congo. At times in human history, social consensus has authorized the practices of foot-binding, genital mutilation, suttee, suicide-bombing, honor killing, and unrestrained consumerism.

These are not ideas imposed on unwilling populations by fanatical autocrats. They have been embraced by "the people," woven into their cultures as well-established social practices, and reproduced generation after generation by the formal and informal socialization practices of, among other institutions, the family and school. The shared values of social consensus can yield abhorrent results.

To make such comparisons and see their implications, of course, one must have an elementary concept of institutions or of society. But the new moral science gives us no such concepts. The social world in all its complexity is simply the sum of individuals motivated by their particular interests in order to sustain or increase their personal well-being. Institutions exist, but generally as vague entities whose legitimacy is measured by the degree to which they increase or decrease personal satisfaction.

The focus on individuals may be partly explained by the Benthamite method itself: "the practice of never reasoning about wholes til they have been resolved into their parts, nor about abstractions until they have been translated into realities."[2] By focusing on the parts—individuals—the new moral scientists ignore the structures of social life and the way they operate.

A Moral Science Unreflective about Power

We may also question whether the new moral science is sufficiently reflective about the dynamics of power in the shaping of discourse. As anyone on the inside of academic life knows, vanity, ego, ambition, intellectual fashion, and political bias play enormous roles in channeling careers and influencing research. They affect who gets funding and from where, who gets published where, whose reputation rises and whose falls. These influences are woven into the pressures applied by professional organizations, departments, colleges and universities, and the granting agencies of large foundations, corporations, and the federal government, none of which are disinterested (though some are more so than others). These external pressures on the guilds of philosophy and the sciences are only the most obvious ways the dynamics of power exert themselves.

More subtle are the workings of power and social control within the guild itself. Every social order is constituted by boundaries that delineate what are understood to be right from wrong, true from false, good from bad, appropriate from inappropriate, sacred from transgressive. This is true at all levels and all spheres of social life, whether in international affairs, nation-states and national cultures, regions and local communities, or particular institutions, such as the family, education, the military, philanthropy, and various arenas of vocation. It is also true in academia—including philosophy and science.

In all spheres of social life, boundaries are not only demarcated but rigorously patrolled. Power and privilege are at stake in every social order. Those who have power and enjoy privilege generally want to keep it, while those who don't have it generally want it. People will talk about "what is true," "what is in the interest of the common good," "what our nation is about"—and, by extension, "what constitutes good science" or "serious philosophy"—but woven into all such claims, in ways that are rarely if ever understood or articulated, are complex and powerful interests oriented toward attaining or preserving advantage, achieving or maintaining purity, and building and keeping reputation. These dynamics are not all that is going on, but they are inextricably part of it.

There was a time when theology claimed a privileged epistemic authority. Its claims to truth were embedded within institutions that could protect the power and advantage of the people making those claims. To contradict its assertions or challenge its authority was an act of transgression that was nearly always met with an act of social control oriented toward both punishing the transgressor and rebuilding the solidarity and the authority of the group in power. Sometimes the social control was symbolic—a declaration of heresy, for example. Other times it was physical, as in the case of witches and heretics who were imprisoned, punished, and even burned at the stake. But even when the violence was symbolic, these actions had real consequences, including ostracism, loss of work, suppression and silencing of dissent, and exclusion from participation in social life.

This is what establishments do. Inside established authorities, there is little reflexivity. There is only truth and error, right and wrong, insider and outsider. However we have evolved as a species, we have not outgrown our requirement to draw boundaries and police them. Nor have we evolved beyond our urge to suppress dissent and oppress dissidents from our circles, even in communities that define themselves as "enlightened," "open," or "liberated." We humans can't help ourselves.

These dynamics are certainly at play in the discourse on the new moral science. They shape "who is in" and "who is out," who gets funded and who doesn't, what constitutes a legitimate or illegitimate line of inquiry and contribution, and ultimately what constitutes truth and error.

This is one way to understand the controversy over Thomas Nagel's 2012 book, *Mind and Cosmos*. Nagel appealed to purposeful natural laws to explain the otherwise improbable appearance of life, consciousness, and morality. Reviewers were apoplectic. Steven Pinker dismissed the book as "the shoddy reasoning of a once-great thinker."[3] Brian Leiter, the proprietor of the blog of record for the professional philosophical community, repeatedly chronicled the controversy of the "formerly reputable philosopher" and loudly worried that "the book is going to have pernicious real-world effects."[4] The philosopher Robert Paul Wolff, a former teacher of Nagel, dismissed central arguments in the book as instances of "philosophical malpractice."[5] "If there were a philosophical Vatican," Simon Blackburn announced in the *New Statesman*, "the book would be a good candidate for going on to the Index."[6]

Reasonable people can disagree about many things. The problem was that Nagel was not merely mistaken in the eyes of his critics, he was apostate.[7] As a secular naturalist, he transgressed a boundary, and what followed was a purity event entailing ridicule and denunciation. As with all purity events, the boundary between insiders and outsiders was clarified and the orthodoxy of the group was reaffirmed.

It is also from this vantage point that we see the distinctiveness of Jonathan Haidt's leadership in encouraging "viewpoint diversity" through his website, Heterodox Academy. The very name of this site and "movement" brings into relief the political problems within academic life generally—not least within the discourse of the new moral science. To be sure, Haidt operates from a position of seniority that is almost unassailable, but he has used this position to courageously challenge establishment conventions. In some respects, Haidt's work is the exception that proves the rule.

The sensationalism of the Nagel affair and the novelty of Heterodox Academy together illustrate the dynamics of power in shaping academic discourse. These dynamics are how all groups maintain social control over their members, and they go on all the time, even though most of the time, social control happens quietly and subtly, far from the public eye.

An Ethics Naïve about Its Own Time

Not least, the new moral science operates in a discursive environment that is larger than the guilds it draws on and is shaped by the cultural, economic, and political dynamics of the modern world. The new moral science seems to be oblivious to how it relates to this larger culture.

Not to put too fine a point on it, the new moral science—its utilitarianism, its pragmatism, its secularism, its nihilism—is of a fabric with the larger ethos of modernity as described by social theorists beginning with Max Weber. This is a world whose dominant feature is instrumental (meaning formal, procedural, or functional) and technical rationality, in which the ends of action are pragmatic and the means are evaluated overwhelmingly in terms of the efficiency of results. If values intervene, they are arbitrary rather than intrinsic to the means and ends of action.

This orientation is not only intellectual but also broadly social, not only philosophical but also broadly functional. It not only defines the terms by which scientists and other intellectuals think and do their work, it orients the world of experts and professionals; most importantly, it spills out further as the dominant form of social organization in all of the contemporary world's most powerful institutions. It is written into the warp and woof of capitalism, the state, the military, law, education, medicine, entertainment, and religion. Instrumental rationality not only organizes action but shapes consciousness and culture. It is precisely because institutional rationality is both ideational and structural that it is so powerful. It is pervasive, even if uneven. It affects everyone who operates in its sway.

Instrumental rationality fosters a world of increasing calculability and predictability, and in so doing, it creates certain freedoms and efficiencies. Yet it also leads to a petrification that tends

to undermine agency and dehumanize both labor and relationships, even the most intimate human bonds.

Weber's chief worry a century ago was that a technical and
instrumental rationality, and the world-picture he saw emerging
from it, were flat, blandly uniform, rigidly standardized, without ideals—just bleak.[8] He didn't anticipate either the limits of
what human beings would tolerate in such environments or their
ability to carve out small areas of personal significance in the
interstices of public life. Nor did he anticipate that people would
sustain meaning and action in life's private spaces. That said, the
cultural logic he observed remains pervasive, especially in public
life. Now, though, as Martin Heidegger, Louis Mumford, Jacques
Ellul, Max Horkheimer, Theodor Adorno, Zygmunt Bauman,
Peter Berger, and others have observed, the logic is reinforced
through the proliferation of technology and a technological consciousness—again, not only ideational but also social.

Instrumental rationality relentlessly deconstructs other
sources of value creation. The result is the weakening of all evaluative perspectives, including its own. It creates a world with little
or no ascertainable grounds for conviction. On a day-to-day level,
the dominance of instrumental and technical rationality in all the
major spheres of contemporary life creates increasing pressure
to relate to the world through economic calculations of utility,
impersonal relations, and expert knowledge.

What is the point?

At the very least, there is a strong "elective affinity" (to use
Weber's phrase) between the new moral science and the structures of thought and practice inherent in a regime defined by
instrumental and technical rationality. There is also an elective
affinity between the nihilism the new moral science leads to and
the nihilism of our late modern entertainment, commercial, and

political culture. In addition, the ethical individualism of the new moral science also shares strong affinities with the market individualism of classical liberal economics. In all three cases, there may not be obvious causal connections, but the cultural logics at play in these affinities gravitate toward each other.

Though this is certainly not its intent, the new moral science provides a highly sophisticated intellectualization for our pervasive regime of instrumental rationality. It offers philosophical legitimation for the cultural and structural dynamics of power at the heart of the contemporary world.

It is important to emphasize that these dynamics are sociological, not psychological. Individual actors may not intend these affinities. In fact, their personal intentions may be oriented in precisely the opposite direction. Intention is irrelevant. What matters here—the point of connection—is the way the moral logic of one (the new moral science) mirrors the cultural logic of the other (the regime of instrumental rationality). It is for this reason that the new moral science can offer no opposition to the ethical dilemmas created by a regime of instrumental and technical rationality.

THE PROMETHEAN TEMPTATION

We could be more expansive in working through how human and social contingencies qualify, compromise, and chasten the claims of science and philosophy to disinterested truth, but we feel the point has been made. An ethics and a moral science that operated with an adequate understanding of history and culture, that was aware of the dynamics of power and its abuse, and that was aware of its relationship with its own culture would be considerably more modest in its claims.

Without such awareness, one is vulnerable to the Promethean temptation to overreach. In this case, it is a temptation to turn science from a method into a metaphysic—from a set of tools, a set of rules, and a discursive orientation into the ground of all being. There is also the temptation to impose on reality a false coherence that is incapable of acknowledging complexity, subtlety, and variation in the natural or social world. As we've seen, some of those working in the new moral science seem inclined to yield to that temptation or at least play at its threshold.

Bearing the Moral Weight of Our Moment

At the same time, we cannot help but be struck by the banality of scientific ethics as currently defined, and we wonder if it can bear the weight of our age's moral concerns. Precisely because the new moral science has no adequate conception of the social, it has no conceptualization of human goods outside of mutually agreeable individual mental states. Public goods have no existence outside of individual subjectivity, no ontology, no transcendence against which the self would be held to account. How, then, within the framework of the new moral science is one to make ethical sense of the threats to our global environment, to liberty from the expanding surveillance capacities of the state, to the economy from fraud in the banking sector, to justice from the corruption of political, civic, and corporate leadership? How is one to make sense of the sacrifice necessary to achieve social justice? The new moral science provides no categories for comprehending, much less addressing, questions of collective moral failure or aspiration.

And, because a moral nihilism underwrites the new moral science, there is nothing intrinsic to it that would set itself against the dehumanizing impulses inherent in the regime of instrumental rationality. Nor is there anything intrinsic to it that would

define itself against economic and social inequality, the exploitation of the weak and the poor, the despoiling of the environment,
the oppression of minorities, or any other depredation. The new
moral science, in itself, provides no resources for either affirming
any moral ideal or resisting any injustice.

What about happiness and well-being? We have observed
already just how historically and culturally tendentious the concept of happiness is in the new moral science, and how poorly it
is operationalized for scientific purposes. The prima facie case
against happiness as a scientifically useful concept is very strong:
people find subjective well-being in as many different sources
as one can imagine. What could be more variable? Middle-
aged introspective academics, Nepalese Sherpas, horny American teenagers, religiously observant Jews, Kalahari bushmen,
Tibetan monks, members of a biker gang, construction workers,
ISIS fighters, and so on, all find happiness in different ways. What
is the happiness they share in common? Except for the word itself,
what common currency of utilitarian exchange could permit us to
resolve conflict and pursue justice? On the face of it, there is no
way to compare one person's happiness to another's, or even to
compare a single person's experience of happiness at one moment
to anything she experienced in the past. There are no units by
which comparison can be made.

People may also find subjective well-being within regimes the
new moral scientists would certainly regard as evil. We know,
for example, that years spent in the *Hitlerjugend* (Hitler Youth)
were among the happiest in the lives of 7.3 million young Nazis.
There they found camaraderie, fellowship, common purpose, and
opportunities to grow, develop, and work together. (Now there
was a social consensus![9])

In the end, happiness, as it is understood today, is a highly subjective category reflecting the historically contingent cultural consensus of late modern Western middle- and upper-middle-class liberalism.[10] It is doubtful that many citizens of Tehran, Soweto, Beijing, or the blighted areas of Detroit, Los Angeles, and Washington, DC, share that consensus. It is dubious whether one could find even a partial social consensus across such differences. But even if one could, there is a genuine question about whether it could resist conflict or sustain passion toward collective moral ideals.

These questions will seem rhetorical, but given the world we live in, it is both justifiable and vital to ask: Would the civil rights movement in the United States have achieved its successes through the kind of utilitarian consensus Joshua Greene speaks of? Would apartheid in South Africa have ended because people worked through the calculations of shared happiness? Could the aspirations of radical Islam be addressed through a conversation about subjective well-being? And could everyday heroism—such as caring for an elderly parent or a severely disabled child over many, many years, risking one's job to blow the whistle on corruption at work, or providing a kidney for a stranger needing a transplant—be achieved and sustained though a rational calculation of cost and benefits? There are good reasons to doubt this.[11]

Against the moral differences at the root of the modern world's conflicts, the new moral science looks weak and unconvincing. Absent an engagement with history, culture, and the humanities, the new moral science appears impoverished, superficial, and trite.

If Not Science, Then What?

As we argued at the beginning of our inquiry, the question "Can science be the foundation of morality?" actually conceals a more profound question—namely, can science rise above all of the differences that divide us and cut through the complexity of knowledge and information that characterizes the modern world to become the foundation of a just and humane social order? Against the epic failure of religion to provide a unifying and peaceable solution to the problems of difference and complexity in the modern world, can science—in effect—provide an alternate magisterium upon which human flourishing could be established? This is the question that we believe has animated the historical quest for over four hundred years, and why, despite the repeated philosophical and empirical failure to make progress toward a solution, the quest will likely continue.

This larger question has significant political consequences. As we noted before, it is no accident that all of the moral scientists of the early modern period—from Grotius and Pufendorf to Hobbes and Locke to Hume to Bentham and Mill—were also well-known and influential political theorists. They understood that the possibilities of a liberalizing political order rested upon the authority of a cultural order that provided basic and shared understandings about the nature of a good and just society. If a political order couldn't rest upon religion, then the only conceivable alternative at that time was "Right Reason," for which science was the final authority. It would have to do what religion had not done and, so far as they could tell, could not do.

But now, if science cannot be the foundational authority of a just and humane moral and political order—if, indeed, the attempt to make science the foundational authority of morality could very well lead to the legitimation of a nihilistic technocratic order—then what is the alternative?

The question of the moral foundations of a good and just society is certainly one of the central philosophical, social, and political puzzles of the modern world since the Reformation. For us to propose an answer in the last pages of the book would be folly. But the urgency for a solution is palpable, made all the more so by the sense that we are in new territory; that our philosophical and political theories and our procedural tools for adjudicating disagreement are proving inadequate to the challenges we now face.

But surely, we cannot settle for looking into the abyss. That would be a different kind of folly.

At this point, the thought occurs to us that the premise of the question might be wrong. Perhaps the problem with both science and early modern religion is that they were attempts to find a single, totalizing, and universal foundation for moral discourse acceptable to all. In both cases, the effort represented different ways of achieving a foundation for a just and humane social order by denying, avoiding, or transcending the knotty, seemingly irreducible problem of difference.

But what if there is no way around the problem of difference? What if the only way to achieve even minimally shared understanding is *through our differences?* This framing of the question is only reinforced by the fact that, historically, there has never been a generic morality capable of distilling all of the varieties of moral understanding and experience. Historically speaking, moralities *only* exist in their particularity and in the particular communities that sustain them.

If this is the case, then the only way we are to sustain the goods that we have achieved in the modern age, and to expand their reach, is to find a common moral understanding *through* our particularities—through our differences—and not in spite of them. Surely there are goods we can all affirm and corruptions we can all repudiate, despite coming from different perspectives.

Needless to say, we are not talking about marginal dissimilarities of taste or lifestyle, but about the fundamental differences that have defined the culture wars for the last half-century. These are differences rooted in opposing visions of national identity and purpose, disparate sources of moral authority, competing ways of life, all rooted in rival understandings of the good, the true, and the beautiful. Working through *those* particularities, *those* differences, would be no small achievement.

Perhaps there is more agreement on the basic understandings about the nature of a good and just society than we think. Perhaps the proceduralism of our liberal political order is more robust than we imagine. These matters are debatable. That said, we do believe that whatever we do share by way of basic moral and cultural understandings is under assault and far from stable.

In such a context, simply making our differences intelligible to one another would be a start. The reason for this, of course, is found in one of the fundamental premises of democracy itself, namely the agreement not to kill each other over our differences, but rather to talk through them. It is in the deepening of the quality of our public discourse on those matters that divide us so profoundly that we have any hope of finding some common ground.

Serious and substantive engagement over such matters would be proto-democratic. It would allow a measure of subtlety and mutual intelligibility rooted in a recognition of our common humanity that is hard to find in our public discourse today. These practices are also consistent with the highest ideals of public education and of the liberal arts in particular.

Absent a more serious and substantive public discourse, the alternative is simply the imposition of one understanding of the good life over all others—which is surely what distinguishes authoritarian regimes over aspirationally liberal and democratic ones.

We are not naïve. Nearly everything in our historical moment would seem to be at war with such possibilities. But at least these appear to be *possibilities*, unlike a successful science of morality.

So, if we are not to succumb to what the Enlightenment *philosophes* commonly described as the "dread and darkness of the mind,"[12] it would seem that we have no choice but to continue the hard work of making sense of the complex, confusing, and conflicted realm of the moral life. In this, science does have an important role to play. After all, there is much about us that does admit of empirical study. But there is finally no substitute for history, literature, poetry, philosophy, sociology, and the world's great religious traditions—no substitute for understanding morality on its own terms—as we struggle to achieve a more just, inclusive, and humane world.

Notes

PREFACE

1. Sam Harris, *The Moral Landscape: How Science Can Determine Human Values* (New York: Free Press, 2010); Michael S. Gazzaniga, *The Ethical Brain: The Science of Our Moral Dilemmas* (New York: Harper-Perennial, 2006); E. O. Wilson, "The Biological Basis of Morality," *The Atlantic*, April 1998; Robert Johnson, *Rational Morality: A Science of Right and Wrong* (Great Britain: Dangerous Little Books, 2013); Paul Zak, *The Moral Molecule: The Source of Love and Prosperity* (New York: Dutton, 2012); Marc D. Hauser, *Moral Minds: How Nature Designed Our Universal Sense of Right and Wrong* (London: Abacus, 2006); Patricia Churchland, *Braintrust: What Neuroscience Tells Us about Morality* (Princeton, NJ: Princeton University Press, 2011); Laurence R. Tancredi, *Hardwired Behavior: What Neuroscience Reveals about Morality* (New York: Cambridge University Press, 2005).

CHAPTER I

1. Robert Wuthnow, *Communities of Discourse* (Cambridge, MA: Harvard University Press, 1989), 2–3.
2. The Reuters Guide to Good Information Strategy, 2000, http://jmab.planetaclix.pt/GesInf/Aula5/The_Reuters_Guide_to_Good_Information_Strategy.pdf.
3. An exabyte is a billion gigabytes.
4. Chad Wellmon, "Why Google Isn't Making Us Stupid . . . or Smart," *The Hedgehog Review* 14, no. 1 (Spring 2012), http://www.iasc-culture.org.
5. Daniel J. Levitin, *The Organized Mind: Thinking Straight in the Age of Information Overload* (New York: Penguin, 2014), 6–7.

6. Richard Wurman, *Information Anxiety* (New York: Doubleday, 1989).

7. Sam Harris, *The Moral Landscape: How Science Can Determine Human Values* (New York: Free Press, 2010), 7.

8. Bertrand Russell, *The Impact of Science on Society* (New York: Simon & Schuster, 1953), 113.

9. See, as just one example, Armond Marie Leroi, "One Republic of Learning: Digitizing the Humanities," *New York Times*, February 13, 2015.

10. Harris, *Moral Landscape*, 7.

11. Yet all too often the questions "Can science teach us anything about morality" and "Can science be the foundation of morality" are conflated. The reasons are obvious enough: if we learn irrefutable facts about the nature of morality from primates or from neuroscience, are we not obligated to follow them? Perhaps—many people believe so. But perhaps not—perhaps we should resist what we learn, particularly if what we discover scientifically is that the moral psychology behind human behavior is largely selfish, dominating, and prone to violence, or that science supports the old social Darwinist idea of the moral inferiority of racial minorities or the "biologically unfit." It depends.

12. This book is about a certain kind of *ethical* naturalism—a version that holds that ethics itself can be explained in the terms of empirical science. As a general phenomenon, ethical naturalism is the view that "ethical facts about such matters as good and bad, right and wrong, are part of a purely natural world—the world studied by the sciences." (See Nicholas Sturgeon, "Ethical Naturalism," in *The Oxford Handbook of Ethical Theory* [New York: Oxford University Press, 2006], 91.) But many philosophers are ethical naturalists who nevertheless deny that science can discover and tell us how to live. Their thought is that while objective human morality is ultimately built up out of the same features of the world that science studies, science itself cannot tell us what morality is or says. So while some ethical naturalists do think science can tell us how to live, not all do.

13. The theoretical and conceptual resources for discourse analysis draw from many, including Erving Goffman (*Frame Analysis: An Essay on the Organization of Experience* [Harmondsworth, England: Penguin, 1974]; "The Interaction Order," *American Sociological Review* 48 [1983]: 1–17); Mikhail Bakhtin ("The Problem of Speech Genre," in *Speech Genre and Other Late Essays* [Austin: University of Texas Press, 1986], 60–102); Michel Foucault (*The Birth of the Clinic: An Archaeology of Medical Perception* [London: Tavistock, 1973]; *Power/Knowledge* [Brighton, England: Harvester, 1984]); and Ernesto Laclau and Chantal Mouffe (*Hegemony and Socialist Strategy* [London: Verso, 1985]). Among the most brilliant empirical studies in discourse analysis is Robert

Wuthnow's *Communities of Discourse: Ideology and Social Structure in the Reformation, the Enlightenment, and European Socialism* (Cambridge, MA: Harvard University Press, 1993).

14. See, for example, http://cushmanlab.fas.harvard.edu/, http://moralitylab. bc.edu/, http://www-bcf.usc.edu/-jessegra/index.html, http://columbiasam-clab.weebly.com/. Others can be found at the University of Iowa, Notre Dame, Penn State, UNC-Chapel Hill, UC-Irvine, and Macalester College. Many of these laboratories operate around the career of a single professor, and they come and go with that professor's mobility.

15. http://scienceofvirtues.org/.

16. http://www.phil.cam.ac.uk/seminars-phil/moral-psychology.

17. http://www.moralpsychology.net/. Out of the twenty-eight scholars in the initial group, they produced over three hundred articles and at least fifteen books since that first meeting, and their work culminated in the Oxford *Moral Psychology Handbook* in 2010.

18. http://moralresearchlab.org/.

19. See, for example, http://www.yourmorals.org/blog/author/jon/; http://scienceblogs.com/pharyngula/2013/02/18/scientific-morality-an-example/; http://www.npr.org/sections/13.7/; http://moralarc.org/can-science-deter mine-moral-values/; http://moralarc.org/counter-refutation-shermer-responds-to-book-reviews/; "The Moral Universe" in *Scientific American*, https://blogs.scientificamerican.com/moral-universe/about/; https://kenan malik.wordpress.com/.

20. As to academic conferences, there are the annual meetings of the Moral Psychology Research Group. There are dozens of others as well: for example, conferences on "The Science of Morality" in London in 2002; "The Psychology and Biology of Morality" at Dartmouth in 2003; "Moral Biology: What (If Anything) Can the Mind Sciences and Evolutionary Biology Tell Us about Law and Morality?" at the Harvard Law School in 2010; "The Evolution of Morality" at Oakland University in Rochester, MI, in 2014; "The Moral Animal: Virtue, Vice & Human Nature," sponsored by the New York Academy of Sciences in 2015. The list goes on.

21. TED talks have been given by many of the authors listed, including Sam Harris, Jonathan Haidt, Steven Pinker, Marc Hauser, Michael Shermer, and Paul Zak. For a different kind of public conference, see https://origins.asu.edu/events/great-debate-can-science-tell-us-right-wrong/. The most important conference for public intellectuals was the conference sponsored by the Edge Foundation; see https://www.edge.org/event/the-new-science-of-morality/.

22. https://www.templeton.org/about/sir-john.

23. John Brockman is an impresario of no small order. Among other things, he

is the literary agent of Haidt, Pinker, Shermer, Hauser, Harris, and Dawkins, among many others. Relatedly, and more importantly, he is the founder of Edge Foundation Inc., an operating foundation whose goal is to remake public culture in light of science. He describes it as an association of science and technology intellectuals created in 1988 as an outgrowth of the "Reality Club," an informal gathering of intellectuals who met from 1981 to 1996 in New York City. Brockman is also the editor and publisher of Edge (www .edge.org), launched in 1996 as the online version of the Reality Club and "as a living document on the Web to display the activities of the so-called 'Third Culture.'"

The mission of the foundation and its activities is to bring together "remarkable people and remarkable minds"—the top inventors, scientists, technicians or "digerati," philosophers, entrepreneurs, journalists, and artists who are at the center of contemporary intellectual, technological, and scientific innovation. For Brockman, these are representative's of the "Third Culture" that, through their work, are "redefining who and what we are." In his own words, the work of the foundation and its activities is to foster "a conversation," a "discourse." Though by Brockman's count, it began with 23 scientists and philosophers, by 2012 edge.org boasted 660 members, and by early 2017, there were 890 who were associating with the portal (see https://www.theguardian.com/technology/2012/jan/08/john-brockman-edge-interview-john-naughton and https://www.edge.org/people/). Day to day, the Edge functions as a symbolic hub for the dynamic of scientists engaging with public life. Its cultural output is impressive: since 2006, there have been more than twelve hundred articles in major news outlets that refer to the Edge in some form or fashion, more than fourteen hundred books published by Edge contributors since its inception.

Operationally, the Edge Foundation and Edge.org sponsor, among other things, master classes with famous intellectuals, seminars (of which the conference on the new science of morality was one), a yearly question, and the annual Billionaires' Dinner, once described by GQ as "row upon row of the world's most successful (and richest) human beings." It includes the presidents/founders of AOL, eBay, Space X, Google, Amazon, Facebook, Yahoo, Microsoft, TED, PayPal, Twitter, Apple, and others.

The Edge website operates as a hub for reframing public culture broadly in the light of science. The discourse on science and morality is just one important node within it. Its importance to our discussion is that it provides a broader plausibility structure for the new science of morality.

24. See Hansfried Kellner and Frank W. Heuberger, *Hidden Technocrats: The New Class and New Capitalism* (New Brunswick, NJ: Transaction Publishers, 1992); James Burnham, *The Managerial Revolution* (New York: John Day, 1941).

25. See, for instance, Kwame Anthony Appiah's discussion in *Experiments in Ethics* (Cambridge, MA: Harvard University Press, 2008), 21–28.

26. *Treatise of Human Nature* 3.1.1. What Hume actually took the problem to be and what the generations since have taken it to be may be two different things, and we are concerned with the latter. The idea that "You can't get an ought from an is" is often asserted as a barrier to any science of morality, and this understanding is what interests us, more so than Hume's intended meaning.

27. As witnessed by the routine attempts to counter or avoid the objection in recent literature. See, for instance, Patricia Churchland, *Braintrust: What Neuroscience Tells Us about Morality* (Princeton, NJ: Princeton University Press, 2011), 4–8; Harris, *Moral Landscape*, 38, 42, 219–20; Paul Thagard, *The Brain and the Meaning of Life* (Princeton, NJ: Princeton University Press, 2010), 183.

28. The Is/Ought Problem is understood in more than one way. The two most common appear to be these:

The Fact/Value version: There are factual claims—claims about what is— and there are evaluative claims—claims about what ought to be—and you can't derive any evaluative claim from factual claims.

The Moral Content version: There are claims that aren't about anything moral—that have no moral or normative content—and those that are about something moral—that do have moral content—and you can't derive any of the latter from just the former. That is, you can't derive claims with moral content from claims lacking moral content.

On closer examination, however, only the Moral Content version here passes muster. The problem with the Fact/Value version concerns its claim that you can't validly infer value statements from factual statements. If this is true, then value statements can't also be factual statements. Otherwise, from the facts "enslaving human beings is wrong" and "Sam is a human being," you could derive the evaluative claim, "It is wrong to enslave Sam," which would be a case of deriving a value claim from factual claims, and so would make this version of the Is/Ought Problem false.

But if value statements aren't supposed to be factual, then how are we to understand them? This turns out to be a big problem for the Fact/Value version. For these value statements that supposedly can't be inferred from facts are either the sorts of things that can be true or false (like propositions or some sentences) or not (like feelings or mere exclamations). They either make some claim about the world that may or may not accurately represent it, or they're nondeclarative statements like, "Look over here," or "Yay for ice cream!" But if they can be true or false, then why aren't they recognized as the kind of things that can be "facts"? Aren't factual statements just true statements? But if the assumption is that they aren't true or false, then it makes this version of the Is/Ought Problem trivially true. After all, statements that aren't true or false can't be derived from any statements, let alone *factual* statements. "Yay for ice cream!" and the like are neither true nor false and hence just aren't the sorts of things that can be the conclusions of a chain of reasoning. So, the Fact/Value version is either false or automatically true in an unhelpful way. Either way, this version of the Is/Ought Problem is an unhelpful tool for evaluating moral thought.

Alternatively, the Moral Content version doesn't frame the Is/Ought Problem in terms of facts and values. Instead, the issue here is whether a claim that has moral or evaluative content can be derived from claims that don't. That is, this version says you can't infer a claim that uses moral ideas from claims that don't use any moral ideas. So on this way of thinking about the Is/Ought Problem, deriving "it is wrong to enslave Sam" from "slavery is wrong" and "Sam is a human being" is not a case of deriving an ought from an is. For the claim that slavery is wrong has moral content—it involves the idea of moral wrongness—and so the derivation would just be a case of deriving an ought from another ought.

Importantly, this version of the Is/Ought Problem would prevent the derivation of claims with moral content from claims lacking moral content, giving sense to the original idea of the Problem: not being able to derive an "ought" from an "is." For instance, suppose someone wanted to derive "It is wrong to enslave Sam" only from the claim that "Sam is a human being." It isn't clear that this works—the premises merely state that Sam is a human being and don't tell us anything about what actions are permitted or forbidden toward someone in virtue of their being human. We would need to supplement "Sam is a human being" with "It is wrong to enslave human beings" to derive the conclusion. And this is exactly what this version of the Is/Ought Problem diagnoses.

Perhaps the Is/Ought Problem is often unpersuasive because it is difficult to argue in general terms that the ethical really is its own autonomous sphere. While there are individual cases where it is evident that science fails in its attempt to draw this or that ethical conclusion from nonethical premises, this doesn't automatically tell us anything about these sorts of inferences as such.

CHAPTER 2

1. That is, found in science as we would recognize it today: a particular practice with specific methods of inquiry. As Peter Harrison points out, what "science" meant in the early modern era was complicated by the fact that the term's meaning was undergoing a transition from an older, medieval notion in terms of a type of internal intellectual virtue to that of an external practice

in the sense just articulated. See his The Territories of Science and Religion, (Chicago: The University of Chicago Press, 2015).

2   Peter Dear, *Revolutionizing the Sciences: European Knowledge and Its Ambitions, 1500–1700* (Princeton, NJ: Princeton University Press, 2001), 3.

3.   Stephen Nadler, "Doctrines of Explanation in Late Scholasticism and in the Mechanical Philosophy," in *The Cambridge History of Seventeenth-Century Philosophy*, ed. D. Garber and M. Ayers (Cambridge: Cambridge University Press, 1998), doi:10.1017/CHOL9780521307635.

4.   Harrison, Territories of Science and Religion, 87–88.

5.   Mark Murphy, "The Natural Law Tradition in Ethics," in *The Stanford Encyclopedia of Philosophy*, Winter 2011 edition, ed. Edward N. Zalta, http://plato. stanford.edu.

6.   Stephen Darwall, "Norm and Normativity," in *The Cambridge History of Eighteenth-Century Philosophy*, ed. Knud Haakonssen (Cambridge: Cambridge University Press, 2006), 989.

7.   Most medieval natural philosophers believed that questions and problems in natural philosophy could not be scientifically demonstrated by reason or experiment. Only probable, plausible, or conjectural responses could be provided for hundreds of questions, such as "whether the celestial spheres are moved by one or several intelligences," "whether the heaven is moved with exertion and fatigue," "whether a comet is a terrestrial vapor," "whether the middle region of air is always cold," or "whether elements remain formally in a compound [or mixed] body." "Aristotelian natural philosophers were unable to formulate decisive answers for most such questions. . . . We have no evidence that they were dissatisfied with multiple responses to a given question or that they believed that better answers had to be found." See Edward Grant, *Foundations of Modern Science in the Middle Ages* (Cambridge: Cambridge University Press, 1996), 158–59.

8.   Diarmaid MacCulloch goes so far as to say that "the Reformation was not caused by social and economic forces, or even by a secular idea like nationalism; it sprang from a big idea about death, salvation, and the afterlife" (*All Things Made New: The Reformation and Its Legacy* [Oxford: Oxford University Press, 2016], 3).

9.   Matthew White, "Selected Death Tolls for Wars, Massacres and Atrocities before the 20th Century," http://necrometrics.com.

10.   See James Tully's introduction to Pufendorf's *On the Duty of Man and Citizen*, ed. Tully (Cambridge: Cambridge University Press, 2003), xix, xx. Schneewind discusses the influence of war on Montaigne and Charron, and later on Hobbes. See his *The Invention of Autonomy* (Cambridge: Cambridge University Press, 1998), 57, 83. Also, it isn't accidental that one of Grotius's primary works is titled "On the Law of War and Peace."

11. Knud Haakonssen, "Early Modern Natural Law," in *The Routledge Companion to Ethics*, ed. John Skorupski (Oxford: Routledge, 2010), 77.

12. For an account of the growing complexity of finance and the emergence of the role of the financier, see Joan DeJean, *How Paris Became Paris* (New York: Bloomsbury, 2014).

13. Prolegomena to Hugo Grotius, *The Rights of War and Peace* (repr. Indianapolis, IN: Liberty Fund, 2005), 1.xviii.

14. Dear, *Revolutionizing the Sciences*, 126.

15. See the introduction in Ian Hunter, *Rival Enlightenments* (Cambridge: Cambridge University Press, 2001), for a detailed description of this impression in early modern Germany.

16. J. D. Bernal, *Science in History* (London: C. A. Watts & Co., 1957), 251–53.

17. Francis Bacon, *The Advancement of Learning*, ed. Joseph Devy (1605; New York: P. F. Collier and Son, 1901), 58.

18. Bacon, *Advancement of Learning*, 42.

19. Dear, *Revolutionizing the Sciences*, 59–61; Steven Shapin, *The Scientific Revolution* (Chicago: University of Chicago Press, 1996), 87–88.

20. See Steven Shapin and Simon Schaffer, *Leviathan and the Air-Pump: Hobbes, Boyle, and the Experimental Life* (Princeton, NJ: Princeton University Press, 1985).

21. This increased focus on mathematics was made possible by the preservation of and commentaries on ancient texts from medieval Muslim scholars, such as Averroes (1126–1198).

22. Shapin, *Scientific Revolution*, 58–59.

23. *Opere Complete di Galileo Galilei* (Florence, 1842), 4:171.

24. Eduard Jan Dijksterhuis, *The Mechanization of the World Picture: Pythagoras to Newton* (Princeton, NJ: Princeton University Press, 1986), 434.

25. Many philosophers in the seventeenth and eighteenth centuries were also physicians. Growing acceptance of corpuscularianism began to contribute to the quality of medical research and treatment. Thinking of the human body as being built up out of stable particles contributed to thinking that the body could be profitably studied, understood, and to some degree, controlled. Richard S. Westfall, *The Construction of Modern Science* (Cambridge University Press, 1971, [chap. 5]).

26. E. A. Burtt, *The Metaphysical Foundations of Modern Science*, rev. ed. (New York: Doubleday Anchor Books, 1954), 303.

27. Rene Descartes, *Le Monde*, chap. 6.

28. Burtt, *Metaphysical Foundations of Modern Science*, 78.

29. The Whiggish view of history continues to be very popular. As Michael Shermer put it, "We can trace the moral arc through science with data from

many different lines of inquiry, all of which demonstrates that in general, as a species, we are becoming increasingly moral. As well, I argue that most of the moral development of the past several centuries has been the result of secular not religious forces, and that the most important of these that emerged from the Age of Reason and the Enlightenment are science and reason, terms that I use in the broadest sense to mean reasoning through a series of arguments and then confirming that the results are true through empirical verification." In *The Moral Arc: How Science and Reason Lead Humanity toward Truth, Justice, and Freedom* (New York: Henry Holt, 2015), 3. See also Steven Pinker, "Science Is Not Your Enemy," *New Republic*, August 6, 2013.

30. Quoted in Peter Gay, *The Enlightenment: An Interpretation, Vol. I. The Rise of Modern Paganism* (New York: Alfred. A. Knopf, 1966), 184.

31. See for instance Helen Hattab's recent investigation of Descartes's arguments against the scholastic doctrine of substantial forms. Not only were Descartes's arguments often secondary to his rhetoric, but, once made clear, were eminently resistible (*Descartes on Forms and Mechanisms* [Cambridge: Cambridge University Press, 2009]).

32. See Rob Iliffe, *Priest of Nature: The Religious Worlds of Isaac Newton (New York: Oxford University Press, 2017).*

33. Shapin, *Scientific Revolution*, 136.

34. Much of what we would today call "science" was at that time known as "natural philosophy." But because our intent is to tell a story to today's readers, we will refer to as "science" those parts of natural philosophy that we would today recognize as scientific, while recognizing that this is anachronistic.

35. See James Tully's introduction to Samuel von Pufendorf's *On the Duty of Man and Citizen*, ed. James Tully (Cambridge: Cambridge University Press, 1991), xvi–xvii. See also Terence Irwin in his *The Development of Ethics*, 3 vols. (New York: Oxford University Press, 2008), 2:122.

36. As Michael Seidler puts it, "Pufendorf and those who shared his outlook claimed a kind of Baconian novelty for their enterprise. This lay in its rejection of the exclusivity of previous authorities and sects, in its dismissal of metaphysics and theology as foundations for philosophy, and in its assertion of an 'eclectic' privilege not only to mix and compare different perspectives, but also to place new bodies of knowledge on a so-called scientific footing" (Seidler, "Pufendorf's Moral and Political Philosophy," *Stanford Encyclopedia of Philosophy* [Spring 2013 Edition], ed. Edward N. Zalta, plato.stanford.edu).

37. Jerome Schneewind, *The Invention of Autonomy* (Cambridge: Cambridge University Press, 1998), 127.

38. As Schneewind puts it in *Invention*, Grotius's approach "takes as central the existence of an enduring tension between our social and our antisocial

dispositions and needs, treats natural laws as empirically discoverable prescriptions for living together despite that tension, and refuses to use a substantial conception of the highest good to derive specific laws" (119–20).

39. Richard Tuck, "Grotius, Carneades, and Hobbes," *Grotiana* 4 (new series): 43–62, 1983; Schneewind, *Invention*, 82.

40. Schneewind, *Invention*, 72.

41. For a brief discussion of Grotius's status as the first to argue for a secular natural law, see Schneewind, *Invention*, 73–74.

42. Grotius, *The Rights of War and Peace*, Preliminary Discourse (Indianapolis: Liberty Fund, Inc., 2005) 40, 110–11.

43. Christian Thomasius, a philosopher and jurist from the generation following Grotius, wrote an early account of this period. "Grotius," he wrote, "was the first to try to resuscitate and purify this most useful science, which had become completely dirtied and corrupted by scholastic filthiness" (Christian Thomasius, *Fundamenta*, preface 1, cited in Schneewind, *Invention*, 66). See also Barbeyrac, *An Historical and Critical Account of the Science of Morality, And the Progress It Has Made in the World, from the Earliest Times down to the Publication of Pufendorf of the Law of Nature and Nations* (London: Printed for J. Walthoe, R. Wilkin, J. and J. Bonwicke, S. Birt, T. Ward, and T. Osborn, 1729), XXIX, 79; Schneewind, *Invention*, 141; Darwall, "Grotius at the Creation of Modern Moral Philosophy," *Archiv fur Gestchichte der Philosophie* 94, no. 3 (October 2012): 320.

44. Schneewind, *Invention*, 78–79.

45. *Rights of War and Peace*, 1.1.4, 138.

46. Grotius "treats [rights] as qualities grounding law, not as derived from law. They are personal possessions, belonging to each of us prior to and independently of our belonging to any community" (Schneewind, *Invention*, 80).

47. Irwin, *Development of Ethics*, 2:126–27.

48. See Jean Hampton, "Hobbes and Ethical Naturalism," *Philosophical Perspectives* 6 (Ethics 1992): 333; Laurence Carlin, *The Empiricists: A Guide for the Perplexed* (London: Continuum, 2009), 43; Schneewind, *Invention of Autonomy*, 88; Jacob Brownowski and Bruce Mazlish, *The Western Intellectual Tradition* (New York: Dorset Press, 1960), 195–96.

49. Haakonssen, "Early Modern Natural Law," 80.

50. Thomas Hobbes, *Leviathan* (Oxford: Clarendon Press, 1909), chap. 6, 41.

51. Hobbes, *Leviathan*, 42.

52. Schneewind, *Invention*, 84; Irwin, *Development of Ethics*, 2: 123.

53. Hobbes, *Leviathan*, 41.

54. John Locke, *An Essay Concerning Human Understanding*, Book II, http://www.earlymoderntexts.com/assets/pdfs/locke1690book2.pdf.

55. G. A. J. Rogers, "Locke, Law, and the Law of Nature," *The Empiricists* (Lanham, MD: Rowman & Littlefield, 1996), 50–51; Haakonssen, "Early Modern Natural Law," 81; Haakonssen, *Natural Law and Moral Philosophy* (Cambridge: Cambridge University Press, 1996), 52, 81.

56. Haakonssen, "Early Modern Natural Law," 81; Schneewind, *Invention*, 157; Irwin, *Development of Ethics*, 2:265.

57. Locke, *Essay Concerning Human Understanding*, 2.30.3–4; 2.31.3–4.

58. Locke, *Essay Concerning Human Understanding*, 3.11.15; 3.11.17–18; 4.12.8; Schneewind, *Invention*, 148.

59. Locke, *Essay Concerning Human Understanding*, 2.28.5.

60. Marquis de Condorcet (Nicolas de Caritat), *Sketch for a Historical Picture of the Progress of the Human Mind*, trans. June Barraclough (London: Weidenfeld and Nicolson [University Microfilms], 1955), 132–33.

61. Denis Diderot, s.v. "Philosophe," *Encyclopedia*, https://archive.org/details/encyclopdieoudio3alemgoog.

62. From Lucretius, along with Cicero, the most quoted figures of antiquity among the *philosophes*. Gay, *Enlightenment*, 103.

63. In 1794 John Adams observed, "The arts and sciences, in general, during the three or four last centuries, have had a regular course of progressive improvement. The invention in mechanic arts, the discoveries in natural philosophy, navigation and commerce, and the advancement of civilization and humanity, have occasioned changes in the condition of the world and the human character which would have astonished the most refined nations of antiquity" ("A Defense of the Constitutions of Government of the United States of America, Against the Attack of M. Turgot in His Letter to Dr. Price, Dated the Twenty-Second Day of March, 1778").

64. Roy Porter, *The Enlightenment* (New York: Palgrave: 2001), 12.

65. Antonia LoLordo points out that the early modern philosophers (those influencing these Enlightenment thinkers) would more likely have understood this sort of view as a form of Epicureanism, but the content of the view is what we would recognize as philosophical naturalism. "Epicureanism and Early Modern Naturalism," *British Journal for the History of Philosophy* 19, no. 4 (2011): 647–64.

66. Condorcet, *Sketch for a Historical Picture of the Progress of the Human Mind*, 132–33 (emphasis added).

67. Charles Frankel, *The Faith of Reason: The Idea of Progress in the French Enlightenment* (New York: King's Crown Press, Columbia University, 1949), 42.

68. Jean-Baptiste le Rond d'Alembert, *Oeuvres* (Paris, 1805), 1.276.

69. Elie Halévy, *The Growth of Philosophic Radicalism* (New York: Macmillan Company, 1928), 6.

CHAPTER 3

1. A key figure here being Hobbes.
2. A key figure here being Samuel Clarke (1675–1729).
3. Julia Driver, "Moral Sense and Sentimentalism," *The Oxford Handbook of the History of Ethics* (Oxford: Oxford University Press, 2013), 358–59.
4. Shaftesbury's major work was the three-volume treatise *Characteristics of Men, Manners, Opinions, Times* (1711). Hutcheson's work: the second essay of *Inquiry Concerning Beauty, Order, Harmony and Design* titled, "Inquiry Concerning Moral Good and Evil" (1725), and *Essay on the Nature and Conduct of the Passions and Affections and Illustrations upon the Moral Sense* (1728). For a good secondary review, see Michael B. Gill, "Ethics and Sentiment: Shaftesbury and Hutcheson," *The Routledge Companion to Ethics* (London: Routledge, 2010), 120–21.
5. Hume, *A Treatise of Human Nature*, ed. David Fate Norton and Mary J. Norton (Oxford: Oxford University Press, 2000), 5.
6. David Hume, *Treatise of Human Nature*, 6.
7. Andrew Janiak, "Newton's Philosophy," *Stanford Encyclopedia of Philosophy* (Summer 2014 edition), ed. Edward N. Zalta, http://plato.stanford.edu.
8. See Norman Kemp Smith, *The Philosophy of David Hume* (London: Macmillan & Co., 1941); Robert Fogelin, *Hume's Skeptical Crisis* (New York: Oxford University Press, 2009).
9. Hume, *Treatise of Human Nature*, 293.
10. For instance, when you think about a table in front of you, what you really know and are aware of is your *perception* of the table, not the table itself. Perhaps your perception of the table is correct, but because humans only experience things through their perceptions, you cannot get beyond your perceptions of the table to check whether your perception really matches up with a table. Or so Hume might say.
11. Here we follow Fogelin's *Hume's Skeptical Crisis*.
12. What makes certain perceptions moral? As Hume had it, there are two kinds of perceptions: impressions and ideas. Impressions are perceptions that "enter with most force and violence," and these include sensations, passions, emotions, and feelings generally. By contrast, ideas are faint copies of the impressions, and ideas are what we use for reasoning and thinking. In this light, Hume argued that since moral thoughts motivate us to action (to act virtuously or to abstain from vicious behaviors), moral thought was a species of impressions—namely, feelings or sentiments. After all, a mere idea cannot motivate us to action—only a sentiment can do that (see *Treatise of Human Nature*, 293–302).
13. On this point Henry Sidgwick remarks, "It would seem that the intellec-

tual energy of this period of English ethical thought had a general tendency to take a psychological rather than a strictly ethical turn. In Hume's case, indeed, the absorption of ethics into psychology is sometimes so complete as to lead him to a confusing use of language" (*Outlines of a History of Ethics* [5th ed., 1902; repr. Indianapolis, IN: Hackett Publishing Company, 1988], 223).

14. Hume, *Treatise of Human Nature*, 471.

15. Hume, *Treatise of Human Nature*, 301.

16. Hume, *Treatise of Human Nature*, 3.1.2, 303.

17. Hume, *Treatise of Human Nature*, 367.

18. Jerome Schneewind, *The Invention of Autonomy* (Cambridge: Cambridge University Press, 1998), 357.

19. Hume, *Treatise of Human Nature*, 368.

20. Hume, *Treatise of Human Nature*, Part 4, Section 4.2.4; Part 2, Sections 8, 10, 11, 12; Part 3.1, respectively.

21. Knud Haakonssen, *The Science of a Legislator* (Cambridge: Cambridge University Press, 1981).

22. For this interpretation of Bentham's repurposing of Hume, we follow Stephen Darwall, "Hume and the Invention of Utilitarianism," in *Hume and Hume's Connexions* (University Park: Penn State University Press, 1994), 58.

23. Jeremy Bentham, *A Fragment on Government*, 1.36n2.

24. French philosopher Claude Helvetius (1715–1771) had argued similarly that all human action could be explained in terms of the pursuit of pleasure and the avoidance of pain and that the principle of utility could play an important role in shaping law and society. See his *Essays on the Mind and Its Several Faculties*. A useful overview of the early utilitarians can be found in Frederick Rosen, *Classical Utilitarianism from Hume to Mill* (London: Routledge, 2015), 95.

25. Jeremy Bentham, *An Introduction to the Principles of Morals and Legislation*, chap. 1. While utilitarians would eventually expand their conception of what constitutes goodness or happiness, pleasure was identified with the good throughout the nineteenth century. See Robert Shaver, "Utilitarianism: Bentham and Rashdall," *The Oxford Handbook of the History of Ethics* (Oxford: Oxford University Press, 2013), 292.

26. For instance, the British legal system at that time had become mired in procedural arcana and bureaucracy and was incapable of providing justice to those seeking it, as Charles Dickens would lampoon a few decades later in *Bleak House*.

27. *The Works of Jeremy Bentham*, published under the Superintendence of His Executor, John Bowring (Edinburgh: William Tait, 1838–1843). eleven vols. http://oll.libertyfund.org/titles/2009#Bentham_0872-01_1090.

28. Julia Driver, "The History of Utilitarianism," *Stanford Encyclopedia of Philosophy* (Winter 2014 edition), ed. Edward N. Zalta, http://plato.stanford.edu.

29. Douglas G. Long, "Science and Secularization in Hume, Smith, and Bentham," in *Religion, Secularization, and Political Thought: Thomas Hobbes to J. S. Mill*, ed. James E. Crimmins (London: Routledge, 1990), 98–99. The quoted phrase from Bentham is from his *Of Laws in General*, ed. H. L. A. Hart (London: Athlone Press, 1970), 120.

30. Jeremy Bentham, *Deontology Together with a Table of the Springs of Action and the Article on Utilitarianism*, ed. J. H. Burns and H. L. A. Hart (London: Athlone Press, 1977), 46–47.

31. Bentham, *Deontology*, 55.

32. Jeremy Bentham, *Theory of Fictions*, included in full in C. K. Ogden, *Bentham's Theory of Fictions* (New York: Harcourt Brace & Company, 1932).

33. This is another instance of ontological pressure—a certain view of what the universe is made of—contributing to a scientific approach to morality. See James Crimmins, *Secular Utilitarianism: Social Science and the Critique of Religion in the Thought of Jeremy Bentham* (New York: Oxford University Press, 1990). It is also worth noting that Bentham, as a utilitarian, was not alone in his scientific aspirations. Helvetius also sought to follow the example of Newton and the new methods of science in the theory of morality. His aim was to "treat morals like any other science and to make an experimental morality like an experimental physics" (*De l'Esprit*, preface). As Charles Frankel put it,

> Helvetius' philosophy gave impetus to the wave of opinion that brought together humanistic and scientific interests. . . . Indeed, it did more: it attempted to show that physical science and the science of human affairs were clearly continuous by providing a "scientific" and mathematical measure of happiness. Helvetius' faith . . . that the ultimate elements to which anything and everything may be reduced are mathematical quantities, suggested the possibility of a calculus of morals based on the assumption that all pleasures are reducible to qualitatively identical units. Pleasures are indistinguishable just as any two physical bodies are ultimately indistinguishable except in terms of their location, their velocity and direction, and their size.

See Frankel, *The Faith of Reason: The Idea of Progress in the French Enlightenment* (New York: King's Crown Press, Columbia University, 1948), 59–60.

34. Elie Halévy, *The Growth of Philosophic Radicalism* (New York: Macmillan Company, 1928), 15.

35. Halévy, *Growth of Philosophic Radicalism*, 32.

36. Halévy, *Growth of Philosophic Radicalism*, 8.
37. David Hartley, *Observations on Man, His Frame, His Duty, and His Expectations* (London: Richardson, 1749), 1.1.2.14 (p. 84).
38. Halévy, *Growth of Philosophic Radicalism*, 13–14.
39. Jeremy Bentham, *The Rationale of Reward* (London: Robert Heward, 1830), 206.
40. Thomas Carlyle (1840), in *On Heroes, Hero-Worship, and the Heroic in History*, ed. M. K. Goldberg, J. J. Brattin, and M. Engel (Berkeley: University of California Press, 1993), 65.
41. John Stuart Mill, *Utilitarianism* (Indianapolis, IN: Hackett Publishing 2001), 7.
42. Henry Sidgwick, *The Methods of Ethics*, 7th ed. (London: Macmillan and Co., 1907), 14.
43. Sidgwick, *Methods of Ethics*, 1.
44. Jerome B. Schneewind, *Sidgwick's Ethics and Victorian Moral Philosophy* (Oxford: Oxford University Press, 1977), 192–93.
45. Schneewind, *Sidgwick's Ethics*, 193.
46. Sidgwick, *Methods of Ethics*, xviii–xix.
47. Alfred Machin, *Darwin's Theory Applied to Mankind*, xvii, as cited in Paul Lawrence Farber, *The Temptations of Evolutionary Ethics* (Berkeley: University of California Press, 1994), 123.
48. William Paley, *Moral and Political Philosophy*, included in *The Works of William Paley* (London: T. Nelson & Sons, 1851), 17.
49. *M Notebook*, MS p. 132e, cited in Robert J. Richards, *Darwin and the Emergence of Evolutionary Theories of Mind and Behavior* (Chicago: University of Chicago Press, 1987), 114.
50. Richards, *Darwin and the Emergence of Evolutionary Theories*, 114–15.
51. Charles Darwin, *The Descent of Man* (1877) in *The Works of Charles Darwin*, ed. P. H. Barrett and R. B. Freeman (London: Pickering and Chatto, 1989), 113.
52. Darwin, *Descent of Man*, 113.
53. Darwin, *Descent of Man*, 113–14.
54. See the very helpful essay by Mario Brandhorst, "Naturalism and the Genealogy of Moral Institutions," *Journal of Nietzsche Studies* 40, no. 1 (Autumn 2010): 5–28. On this point, Darwin wrote, "An instinct to support and aid other members of the group will therefore be reinforced by natural selection. The basic thought is that the members of a group in which social instincts have come to prevail are more likely to survive and rear more offspring than the members of competing, less cohesive groups. Groups in which selfishness is curbed will therefore have an evolutionary advantage over groups that are divided and disrupted by persistent selfish action. If the members of a

more cohesive group are in fact more likely to rear offspring, then the social instincts will eventually prevail" (Darwin, *Descent of Man*, 137).

55. Richards, *Darwin and the Emergence of Evolutionary Theories*, 218. The quotes from Darwin are from *Descent of Man*, 97, 98.

56. Richards, *Darwin and the Emergence of Evolutionary Theories*, 234–35.

57. Darwin, *Descent of Man*, 92.

58. Paul Thompson, Editor's Introduction, *Issues in Evolutionary Ethics* (Albany: State University of New York Press, 1995), 7.

59. Leslie Stephen, *The Science of Ethics*, 217, referenced in Farber, *Temptations of Evolutionary Ethics*, 32.

60. Farber, *Temptations of Evolutionary Ethics*, 32.

61. Farber, *Temptations of Evolutionary Ethics*, 35.

62. Farber, *Temptations of Evolutionary Ethics*, 49.

63. Herbert Spencer, *The Principles of Ethics*, vol. 1 (New York: D. Appleton and Co., 1892), 57.

64. Camilla Kronqvist, "The Relativity of Westermarck's Moral Relativism," in *Westermarck*, Occasional Paper No. 44 of the Royal Anthropological Institute, ed. David Shankland (Herefordchire, England: Sean Kingston Publishing, 2014).

65. In one of many examples, Edward Westermarck discusses a tribe in Fiji:

> In Fiji, also, it was regarded as a sign of filial affection to put an aged parent to death. . . . One reason why the old Fijian so eagerly desired to escape extreme infirmity was perhaps "the contempt which attaches to physical weakness among a nation of warriors, and the wrongs and insults which await those who are no longer able to protect themselves"; but another, and as it seems more potent, motive was the belief that persons enter upon the delights of the future life with the same faculties, mental and physical, as they possess at the hour of death, and that the spiritual life thus commences where the corporeal existence terminates. "With these views," says Dr. Hale, "it is natural that they should desire to pass through this change before their mental and bodily powers are so enfeebled by age as to deprive them of their capacity for enjoyment." (*The Origin and Development of the Moral Ideas* [London: Macmillan and Co., 1906], 389–90.)

66. Maurice Bloch, "Westermarck's Theory of Morality in His and Our Time," in Shankland, *Westermarck*.

67. Farber, *Temptations of Evolutionary Ethics*, 98.

68. Farber, *Temptations of Evolutionary Ethics*, 119.

69. Alfred Russel Wallace, *Contributions to the Theory of Natural Selection* (New York: Macmillan and Co., 1870), 355.

70. St. George Jackson Mivart, *Genesis of Species* (New York: Appleton, 1871), 220–22.

71. Theodore de Laguna, "Stages of the Discussion of Evolutionary Ethics," *Philosophical Review* 14, no. 5 (1905):, 583.

72. Sidgwick, *Methods of Ethics*, v–vi. See also T. H. Green, *Prolegomena to Ethics*, ed. A. C. Bradley (Oxford: Oxford University Press, 1883), and William Ritchie Sorley, *The Ethics of Naturalism: A Criticism* (Edinburgh: William Blackwood and Sons, 1904), for other cogent philosophical critiques of the effort to establish ethics within evolutionary theory. See also Farber, *Temptations of Evolutionary Ethics*, 98.

73. T. H. Huxley, "Evolution and Ethics," *Evolution and Ethics and Other Essays* (New York: D. Appleton and Company, 1905), 80.

74. G. E. Moore, *Principia Ethica* (Cambridge: Cambridge University Press, 1903).

75. Thomas Hurka, "Moore's Moral Philosophy," *Stanford Encyclopedia of Philosophy* (Summer 2010 edition), ed. Edward N. Zalta, http://plato.stanford.edu.

76. Alan Donagan, "Twentieth-Century Anglo-American Ethics," in *A History of Western Ethics*, ed. Lawrence C. Becker and Charlotte B. Becker (New York: Routledge, 2003), 144–45.

77. John Herman Randall, Jr., *The Making of the Modern Mind: A Survey of the Intellectual Background of the Present Age* (Boston: Hoghton Mifflin Co., 1926), 497.

78. Randall, *Making of the Modern Mind*, 498.

79. John B. Watson, "Psychology as the Behaviorist Views It," *Psychological Review* 20, no. 2 (1913): 158–77.

80. John B. Watson, *Behavior: An Introduction to Comparative Psychology* (New York: Henry Holt, 1914).

81. Richards, *Darwin and the Emergence of Evolutionary Theories*, 503–9.

82. James Tully's introduction to Pufendorf's *On the Duty of Man and Citizen*, ed. James Tully (Cambridge: Cambridge University Press, 1991).

83. Michel de Montaigne, *Essays*, 2.12, https://www.gutenberg.org.

84. The unreliability of our mental faculties (in perception and reasoning), the multiple contradictions in our own thinking, and the failure to apply the rules of reason and science reliably mean that our conclusions can never be regarded as certain or foolproof. "All knowledge," Hume famously argued, "degenerates into probability" (*Treatise*, 1.4.1). What is more, the character of causal reasoning relies upon prior observation as evidence for judgments made upon similar circumstances in the present. Ironically, this leads to a certain circularity in our logic and deductions. This uncertainty built into human knowledge and understanding invariably leads, as Hume put it, to "a total extinction of belief and evidence." As Hume wrote in the *Treatise*, "The

intense view of these manifold contradictions and imperfections in human reason has so wrought upon me, and heated my brain, that I am ready to reject all belief and reasoning, and can look upon no opinion even as more probable or likely than another" (1.4).

85. "Letter to Frances Hutcheson," in John Hill Burton, *Life and Correspondence of David Hume* (Edinburgh: William Tait, 1846), 112–13.

86. This points to another sort of skepticism in Hume: disbelief in the relevance or even reality of the subjective self. This is significant because in dismissing the validity of subjective appearances Hume relegates the role of moral agency in moral life. For if the appearance of rationally compelling moral phenomena is illusory, as Hume's view seems to entail, then reflecting on and acting in light of these appearances would appear misguided. We discuss this sort of skepticism and its metaphysical consequences in a bit more depth in chapter 8.

87. A more fitting metaphor for Hume's moral theory might be the relationship between what is experienced in a virtual reality simulator and the digital processing that makes it possible. Sure, it *looks* like you are flying over the Grand Canyon on condor's wings, but really, it's all ones and zeros.

CHAPTER 4

1. For one, there was G. E. M. Anscombe's work revitalizing Aristotelian virtue ethics, e.g., her essay, "Modern Moral Philosophy," *Philosophy* 33, no. 124 (January 1958): 1–19.

2. John Rawls, *A Theory of Justice* (Cambridge, MA: Belknap Press of Harvard University Press, 1971).

3. Extending a line of thought begun a few decades earlier by Jean Piaget, Kohlberg argued that moral development was a form of rational development. Kohlberg thought that as human beings grow and mature, we progress through various stages of moral development, from blindly obeying authoritarian rules until finally reaching a level where we perceive morality to be justified by abstract principles. Don Locke, "A Psychologist among the Philosophers: Philosophical Aspects of Kohlberg's Theories," in *Lawrence Kohlberg: Consensus and Controversy*, ed. Sohan Modgil and Celia Modgil (London: Routledge, 2011), 25.

4. This puzzle—the inability to account for the presence of altruism—was yet another reason why evolutionary accounts of ethics fell out of favor in the first half of the twentieth century.

5. This way of putting the answer is taken from Wilson's later work: David Sloan Wilson and Edward O. Wilson, "Rethinking the Theoretical Foundation of Sociobiology," *Quarterly Review of Biology* 82, no. 4 (December 2007): 345.

6. E. O. Wilson, *Sociobiology: The New Synthesis* (Cambridge, MA: Belknap Press of Harvard University Press, 1975), 562.

7. Wilson, *Sociobiology*, 563. Jonathan Haidt put Wilson's point more succinctly: "Evolution shaped human brains to have structures that enable us to experience moral emotions. These emotional reactions provide the basis for intuitions about right and wrong, and we (or, at least, many moral theorists) make up grand theories afterward to justify our intuitions" (Haidt, "Morality," *Perspectives on Psychological Science* 3, no. 1 (2008): 68.

8. Philip Kitcher, *Vaulting Ambition* (Cambridge, MA: MIT Press, [1987], 184).

9. Evolutionary psychologists today sometimes try to draw a sharp distinction between evolutionary psychology and sociobiology. But the distinctions seem artificial as evolutionary psychology obviously carries on the Wilsonian program of seeking to understand the function of the brain in terms of neurochemistry and evolutionary development.

10. Haidt, "Morality," 68.

11. See, for instance, Frans de Waal, *Good Natured: The Origins of Right and Wrong in Humans and Other Animals* (Cambridge, MA: Harvard University Press, 1996), and Christopher Boehm, *Moral Origins: The Evolution of Virtue, Altruism, and Shame* (New York: Basic Books, 2012).

12. Nikolas Rose and Joelle M. Abi-Rached, *Neuro: The New Brain Sciences and the Management of the Mind* (Princeton, NJ: Princeton University Press, 2013), 38.

13. For instance, moral psychologist Joshua Greene draws crucially on recent work in experimental psychology and cognitive neuroscience. He and others have performed numerous experiments in order to better understand what happens at the neural level when human beings consider moral issues. Greene writes, "This book . . . builds on my own research in the new field of moral cognition, which applies the methods of experimental psychology and cognitive neuroscience to illuminate the structure of moral thinking. Finally, this book draws on the work of hundreds of social scientists who've learned amazing things about how we make decisions and how our choices are shaped by culture and biology. This book is my attempt to put it all together, to turn this new scientific self-knowledge into a practical philosophy that can help us solve our biggest problems" (*Moral Tribes: Emotion, Reason, and the Gap between Us and Them* [New York: Penguin Press, 2013], 5).

14. Jonathan Haidt, *The Righteous Mind* (New York: Vintage Books, 2012), 38. For clear instances of work in this new synthesis, see Boehm, *Moral Origins*, and Robert Wright, *The Moral Animal: The New Science of Evolutionary Psychology* (New York: Vintage Books, 1994).

15. Haidt, *Righteous Mind*, 29.

16. Haidt, *Righteous Mind*, 38. Quoted text within Haidt's quote is from Wilson, *Sociobiology*.

17. For instance, while Joshua Greene and Philip Kitcher do see a significant place for the moral emotions in morality, both ultimately see reason as primarily explanatory. Greene thinks conscious reasoning should manage emotional moral intuition, and Kitcher thinks that morality is ultimately constituted by social problem-solving, which is a rational process. But Greene at least is firmly committed to the primacy of the scientific study of the mind. Greene, *Moral Tribes*; Philip Kitcher, *The Ethical Project* (Cambridge, MA: Harvard University Press, 2011), esp. 96–103.

18. Owen Flanagan, *The Really Hard Problem* (Cambridge, MA: MIT Press, 2007), 46–47.

19. Paul Thagard, *The Brain and the Meaning of Life* (Princeton, NJ: Princeton University Press, 2010), 185.

20. Flanagan, *Really Hard Problem*, 46–47.

21. Thagard, *Brain and the Meaning of Life*, 188. While Thagard puts tremendous emphasis on the empirical study of the brain for understanding morality, he is more explicit than most in that he sees the role of empirical study here as limited. As he puts it, "I don't think that evidence about the brain is by itself sufficient to direct us to any one ethical theory that we ought to adopt, but I will try to show that such evidence puts some constraints on the evaluation of ethical theories" (Thagard, *Brain and the Meaning of Life*, 195).

22. Frans de Waal, *The Bonobo and The Atheist* (New York: W. W. Norton and Company, 2013), 228.

23. This is the view of Greene, *Moral Tribes*, chapter 3.

24. Patricia Churchland, *Braintrust: What Neuroscience Tells Us about Morality* (Princeton, NJ: Princeton University Press, 2011), 13.

25. Flanagan, *Really Hard Problem*, 45.

26. Traditionally, a consequentialist moral theory counts as utilitarian only if the consequences are evaluated as good or bad in terms of pleasure and pain, as was the case with Bentham. Harris (and others advocating consequentialist scientific approaches) broaden their theory to include human happiness and well--being in evaluating consequences. But since utilitarianism is a more recognizable term, and conveys the basic idea better to a nonspecialist audience, we will count Harris under the banner of utilitarianism. Greene and Haidt make the same move—calling happiness-based consequentialism "utilitarianism" (Greene, *Moral Tribes*, 202–4; Haidt, *Righteous Mind*, 314–16). To avoid confusion in the literature, we follow suit.

27. Churchland, *Braintrust*, 8–9. Elsewhere she restated the point: "Morality can be—and I argue, *is*—grounded in our biology, in our capacity for compassion and our ability to learn and figure things out. As a matter of actual

fact, some social practices are better than others, some institutions are worse than others, and genuine assessments can be made against the standard of how well or poorly they serve human well-being" (*Braintrust*, 200).

28. Flanagan, *Really Hard Problem*, 121.

29. Flanagan, *Really Hard Problem*, 124.

30. Philip Kitcher, "Naturalistic Ethics without Fallacies," *Preludes to Pragmatism: Toward a Reconstruction of Philosophy* (New York: Oxford University Press, 2012), 315.

31. Philip Kitcher, *The Ethical Project*, 262.

32. "Eleven Dogmas of Analytic Philosophy," *Psychology Today*, December 12, 2014.

33. Paul Thagard, "Nihilism, Skepticism, and Philosophical Method: A Response to Landau on Coherence and the Meaning of Life," *Philosophical Psychology* 26, no. 4 (2013): 621. This is a pithier summary of the method he presents in Thagard, *Brain and the Meaning of Life*, 211.

34. Thagard, *Brain and the Meaning of Life*, 207. He writes, "Actions have consequences that affect the needs of people; an action is right to the extent that it furthers those needs, and wrong to the extent that it damages them."

35. The new moral scientists do not claim that the majority settles which values will be pursued. Rather, what we mean here is that they appear to select as goal-values candidates that will meet with little resistance from their intended audiences of educated, liberal, Western, English-speakers (e.g., subjective happiness with a few caveats for objective components, like the flourishing of one's loved ones.) Joshua Greene, Sam Harris, and Owen Flanagan, among others, select such values.

36. Kitcher, *Ethical Project*, 288.

37. It is commonly held that naturalism comes in two sorts: ontological and methodological. The ontological version is about what reality is made out of (e.g., physical stuff(. For instance, Alex Rosenberg puts it like this: "The physical facts fix all the facts" (*The Atheist's Guide to Reality: Enjoying Life Without Illusions* [New York: W. W. Norton and Company, 2011]). The methodological version is about the best way to study reality (e.g., empirical science). For instance, Brian Leiter describes it as, "What there is and what we know are questions reliably answered by the methods of the empirical sciences" ("Normativity for Naturalists," forthcoming in Philosophical Issues 25, no. 1). We combine these versions here, since the new synthesis naturalists generally endorse both sorts.

38. The reach of the sciences, from this vantage point, is unlimited. As Owen Flanagan argued, "The sciences can explain, in principle, the nature and the function of art, science, ethics, religion, and politics" (*Really Hard Problem*, 21). It reaches to our understanding of the good: "If one adopts the

perspective of the philosophical naturalist and engages in realistic empirical appraisal of our natures and prospects, we have chances for learning what methods might reliably contribute to human flourishing. This is eudaimonics" (*Really Hard Problem*, 4).

39. Greene explains one especially rigorous way of being a philosophical naturalist in an essay of advice to his fellow neuroscientists working in social psychology: "What, then, are we really trying to do? . . . [T]he mission of social neuroscience, as the offspring of social psychology and neuroscience, is to understand all of human subjective experience in physical terms" ("Social Neuroscience and the Soul's Last Stand," in *Social Neuroscience: Toward Understanding the Underpinnings of the Social Mind*, ed. A. Todorov, S. Fiske, and D. Prentice [New York: Oxford University Press, 2011], 19–20; also available at http://www.wjh.harvard.edu/-jgreene/GreeneWJH/Greene-Last-Stand .pdf). Citations here are from the online version.

Kitcher is not atypical when he argues that the naturalism he endorses "consists in refusing to introduce mysterious entities—'spooks'—to explain the origin, evolution, and progress of ethical practice" (*Ethical Project,* 3). He claims that his approach relies on many "rigorous forms of inquiry" besides the hard sciences, provided that these methods have the resources to "convince current investigators" that they are offering an improved picture of the world. His version of pragmatic naturalism has a place for philosophy, for example, so long as it "live[s] up to the standards our most rigorous investigations set for themselves," and follows the rule "there are to be no spooks" ("Naturalistic Ethics without Fallacies," 304). For Kitcher, it is the sciences that provide the standards of rigorous investigation. So Kitcher retains a place for modes of inquiry other than science, but they can participate only insofar as they achieve the levels of rigor and evidence exemplified by science.

40. Julia Tanney, "Gilbert Ryle," *Stanford Encyclopedia of Philosophy* (Winter 2014 edition), ed. Edward N. Zalta, http//plato.stanford.edu; Howard Robinson, "Dualism," *Stanford Encyclopedia of Philosophy* (Winter 2012 edition), ed. Edward N. Zalta, http://plato.stanford.edu; George A. Miller, "The Cognitive Revolution: A Historical Perspective," *TRENDS in Cognitive Sciences* 7, no. 3 (March 2003): 141–44.

41. Note, for instance, a recent conference featuring several of the new moral scientists called "Moving Naturalism Forward." The sentiment appears to be that naturalism does not go without saying but is imminently worth advocating.

42. See Tamler Sommers's interview with Michael Ruse in Sommers's *A Very Bad Wizard: Morality behind the Curtain* (San Francisco: McSweeney's, 2009); Fiery Cushman, "Morality: Don't Be Afraid—Science Can Make Us Better," *New Scientist* no. 2782 (October 13, 2010); Christopher Boehm, Christopher

Boehm, *Moral Origins*; Tamler Sommers and Alex Rosenberg, "Darwin's Nihilistic Idea: Evolution and the Meaninglessness of Life," *Biology and Philosophy* 18, no. 5 (November 2003): 653–68; Stephen G. Morris, *Science and the End of Ethics* (New York: Palgrave Macmillan, 2015); Mark Johnson, *Morality for Humans* (Chicago: University of Chicago Press, 2014).

43. Greene, *Moral Tribes*, 373–74n.
44. Greene, *Moral Tribes*, 188.
45. Greene, *Moral Tribes*, 172.

CHAPTER 5

1. In philosophy, this is called the "demarcation problem." See Massimo Pigliucci and Maarten Boudry (eds.), *Philosophy of Pseudoscience: Reconsidering the Demarcation Problem* (Chicago: University of Chicago Press, 2013).
2. Harris writes, "Some people . . . [define] science in exceedingly narrow terms, as though it were synonymous with mathematical modeling or immediate access to experimental data. However, this is to mistake science for a few of its tools. Science simply represents our best effort to understand what is going on in this universe, and the boundaries between it and the rest of rational thought cannot always be drawn" (*The Moral Landscape: How Science Can Determine Human Values* [New York: Free Press, 2010], 29).

   He repeats the argument later in the book: "I do not intend to make a hard distinction between 'science' and other intellectual contexts in which we discuss 'facts'—e.g., history. For instance, it is a fact that John F. Kennedy was assassinated. Facts of this kind fall within the context of 'science,' broadly construed as our best effort to form a rational account of empirical reality" (Harris, *Moral Landscape*, 211).
3. These observations were made by Pinker at "The Great Debate: Can Science Tell Us Right from Wrong?" November 6, 2010, Arizona State University.
4. Thagard elaborates:

   First, explanations in science employ detailed mechanisms, which are descriptions of systems of interconnected parts that produce regular changes. . . . Second, science often uses mathematics in its formulation of hypothesis and explanations that connect them with observations. . . . Third, the social structures of science enforce the logical prescriptions of inference to the best explanation more stringently than is found in everyday life. . . . Fourth, scientists are trained not to focus on just those observations that fit with their biases, but rather to conduct systematic observations that collect broad and representative samples of relevant data. . . . Fifth, whereas ordinary people gain evidence only from their senses such as sight, scientists use instruments to observe things and events that are out of reach of direct sense experience. . . . The sixth and probably most important way in which evidence-based inference in science differs from everyday life is the use of experiments (*The Brain and the Meaning of Life* [Princeton, NJ: Princeton University Press, 2010], 23–25).

5. Harris, *Moral Landscape*, 28, 80, 219n19.
6. We searched the journals for articles whose titles or abstracts used any of about twenty terms topically related to ethics or morality. These terms included: "moral," "morality," "good," "bad," "care," "harm," "altruism," "ethics," "ethical," "right," "wrong," "norms," "normative," "selfishness," "empathy," "justice," "fairness," "loyalty," "sympathy," "helpfulness," "cooperation," and "self-interest."

Among general science journals, we looked at *Nature, Science,* and *Proceedings of the National Academy of Sciences.* For social psychology, we looked at *Personality and Social Psychology Review, Advances in Experimental Social Psychology, Journal of Personality and Social Psychology, European Journal of Personality, Journal of Personality, Personality and Social Psychology Bulletin, Social Psychological and Personality Science,* and *Journal of Experimental Social Psychology.* For evolutionary biology, we looked at *Trends in Ecology and Evolution; Systematic Biology; Annual Review of Ecology, Evolution, and Systematics; Molecular Biology and Evolution; Molecular Ecology; Evolution; American Naturalist;* and *Journal of Evolutionary Biology.* For neuroscience, we looked at *Nature Reviews Neuroscience, Trends in Cognitive Sciences, Behavioral and Brain Sciences, Annual Review of Neuroscience, Nature Neuroscience, Neuron,* and *Journal of Neuroscience.* For evolutionary psychology, we looked at *Evolution and Human Behavior, Human Nature,* and *Evolutionary Psychology.* And for primatology, we looked at *American Journal of Primatology, International Journal of Primatology, Primates,* and *American Journal of Physical Anthropology.*

7. Two notable exceptions in our investigations are the works of evolutionary biologist Martin Nowak and neuroeconomist Ernst Fehr. Both are among the most highly cited researchers in the science of morality, considered broadly. However, we chose not to focus on Nowak's work because it appears to be extremely controversial within his field. We wanted to ensure that whoever we engaged with couldn't be dismissed as unrepresentative of his or her broader field. We chose not to engage with Fehr's work because it wasn't clear to us that he attempts to draw any conclusions for moral theory from his research. While he researches the biological analogues of some moral terms—"altruism," "selfishness," etc.—he seems to generally confine himself to the specifically scientific definitions of these terms, definitions that do not clearly have anything interesting to do with their moral analogues.

8. These initiatives were spearheaded by R. A. Fisher, J. B. S. Haldane, George R. Price, and W. D. Hamilton between 1930 and 1964.

9. Robert L. Trivers, "The Evolution of Reciprocal Altruism," *Quarterly Review of Biology* 46, no. 1 (1971): 35–57.

10. For numerous examples, see Frans B. M. de Waal, "Putting the Altruism Back into Altruism: The Evolution of Empathy," *Annual Review of Psychology* 59 (2008): 279–300, doi:10.1146/annurev.psych.59.103006.093625. See also Samir Okasha, "Biological Altruism," *Stanford Encyclopedia of Philosophy* (Fall 2013 edition), ed. Edward N. Zalta, http://plato.stanford.edu.

11. See, for example, Kevin R. Foster, Tom Wenseleers, and Francis L. W. Ratnieks, "Kin Selection Is the Key to Altruism," *Trends in Ecology & Evolution* 21, no. 2 (February 2006): 57–60, doi:10.1016/j.tree.2005.11.020, as well as William O. H. Hughes, Benjamin P. Oldroyd, Madeleine Beekman, and Francis L. W. Ratnieks, "Ancestral Monogamy Shows Kin Selection Is Key to the Evolution of Eusociality," *Science* 320, no. 5880 (May 30, 2008): 1213–16, doi:10.1126/science.1156108. Also, see the evidence cited in the five letters (signed by 130 evolutionary biologists) to *Nature* published in response to the high-profile attack on inclusive fitness by Nowak, Tarnita, and Wilson: *Nature* 471, no. 7339 (March 24, 2011): E1–4.

12. Why Evolution Is True, "New Paper Shows That Nowak et al. Were Wrong: Kin Selection Remains a Valuable Concept in Evolutionary Biology," March 27, 2015, https://whyevolutionistrue.wordpress.com.

13. The behavioral tendency toward biological altruism is almost certainly not located in a single gene or small set of genes but emerges from the genome in complex ways that are presently unknown.

14. For a broad survey of such findings, see John Tooby and Leda Cosmides, "Conceptual Foundations of Evolutionary Psychology," *Handbook of Evolutionary Psychology*, ed. D. M. Buss (Hoboken, NJ: Wiley, 2005); Jaime Confer et al., "Evolutionary Psychology: Controversies, Questions, Prospects, and Limitations," *American Psychologist* 65, no. 2 (2010): 110–26.

15. de Waal, "Putting the Altruism Back into Altruism."

16. de Waal, "Putting the Altruism Back into Altruism," 292

17. *The Primate Mind* (Cambridge, MA: Harvard University Press, 2012), 125. "Emotional responses to displays of emotion in others are so commonplace in animals (de Waal 2003, Plutchik 1987, Preston & de Waal 2002b) that Darwin (1982 [1871, p. 77]) already noted that 'many animals certainly sympathize with each other's distress or danger.' For example, rats and pigeons display distress in response to perceived distress in a conspecific, and temporarily inhibit conditioned behavior if it causes pain responses in others (Church 1959, Watanabe & Ono 1986)."

18. de Waal, "Putting the Altruism Back into Altruism," 283, citing Eisenberg 2000, p. 677.

19. de Waal, "Putting the Altruism Back into Altruism," 285, citing de Waal & van Roosmalen, 1979.

20. de Waal, "Putting the Altruism Back into Altruism," 285. He adds detail:

> A mother ape who returns to a whimpering youngster to help it from one tree to the next—by swaying her own tree toward the one the youngster is trapped in and then drapes her body between both trees—goes beyond mere concern for the other. Her response likely involves emotional contagion (i.e., mother apes often briefly whimper themselves when they hear their offspring do so), but adds assessment of the specific reason for the other's distress and the other's goals.

21. de Waal, "Putting the Altruism Back into Altruism," citing S. D. Preston and Frans B. M. de Waal, "Empathy: Its Ultimate and Proximate Bases." *Behavioral and Brain Sciences* 25, no. 1 (2002): 1–72.

22. "Evidence for altruism based on empathic perspective-taking mostly consists of striking anecdotes, which are admittedly open to multiple interpretations." But de Waal is optimistic: "Given the overwhelming observational evidence for spontaneous helping and cooperation among primates, it seems only a matter of time until other-regarding preferences will be experimentally confirmed" (de Waal, "Putting the Altruism Back into Altruism," 289–290).

23. "Fortunately, with regard to primate altruism, we do not need to rely on qualitative accounts as there exists ample systematic data, such as a rich literature on support in aggressive contexts (Harcourt & de Waal 1992), cooperation (Kappeler & van Schaik 2006), and food-sharing (Feistner & Mcgrew 1989)." (de Waal, "Putting the Altruism Back into Altruism," 289.)

24. de Waal, "Putting the Altruism Back into Altruism," 292.

25. Sarah F. Brosnan and Frans B. M. de Waal, "Fairness in Animals: Where to from Here?" *Social Justice Research* 25, no. 3 (September 5, 2012): 336–51, doi:10.1007/s11211-012-0165-8.

26. Sarah F. Brosnan and Frans B. M. de Waal, "Monkeys Reject Equal Pay," *Nature* 425 (September 18, 2003): 297–99.

27. *Moral Tribes: Emotion, Reason, and the Gap between Us and Them* (New York: Penguin Press, 2013), 120–21.

28. "In response to the footbridge dilemma and others like it, manual mode advises us to maximize the number of lives saved while our gut reactions tell us to do the opposite. The parts of the brain that support the utilitarian answer, most notably the DLFPC, are the same parts of the brain that enable

us to behave flexibly in other domains. . . . And the parts of the brain that work against the utilitarian answer in moral dilemmas, most notably the amygdala and the VMPFC, are the parts of the brain that inflexibly respond with heightened vigilance to things like the faces of out-group members" (Greene, *Moral Tribes*, 172–73).

29. Joshua D. Greene, "Beyond Point-and-Shoot Morality: Why Cognitive (Neuro)Science Matters for Ethics," *Ethics* 124, no. 4 (2014): 695–726.

30. Consequentialism is an ethical view according to which what makes actions right or wrong is just their consequences. Utilitarianism is one species of consequentialism, adding that the consequences that matter for right and wrong are those that promote or undermine overall happiness or pleasure.

31. A. M. Isen and P. F. Levin, "Effect of Feeling Good on Helping: Cookies and Kindness," *Journal of Personality and Social Psychology* 21 (1972): 384–88.

32. K. E. Mathews and L. K. Cannon, "Environmental Noise Level as a Determinant of Helping Behavior," *Journal of Personality and Social Psychology* 32 (1975): 571–77.

33. John Doris and Stephen Stich, "Moral Psychology: Empirical Approaches," *Stanford Encyclopedia of Philosophy* (Fall 2014 edition), ed. Edward N. Zalta, http://plato.stanford.edu.

34. Jonathan Haidt, *The Righteous Mind: Why Good People Are Divided by Politics and Religion*, reprint ed. (New York: Vintage, 2013); Jesse Graham, Jonathan Haidt, Sena Koleva, Matt Motyl, Ravi Iyer, Sean P. Wojcik, and Peter H. Ditto, "Moral Foundations Theory: The Pragmatic Validity of Moral Pluralism," SSRN Scholarly Paper, Social Science Research Network, Rochester, New York, November 28, 2012, http://papers.ssrn.com/abstract=2184440.

35. Graham et al., "Moral Foundations Theory," 7.

36. Haidt, *Righteous Mind*, 145.

37. See Haidt, *Righteous Mind*, chap. 8.

38. Graham et al., "Moral Foundations Theory," 37–38.

39. See section 3 in Graham et al., "Moral Foundations Theory."

40. Jonathan Haidt and Jesse Graham, "When Morality Opposes Justice: Conservatives Have Moral Intuitions That Liberals May Not Recognize," *Social Justice Research* 20, no. 1 (May 23, 2007): 98–116, doi:10.1007/s11211-007-0034-z.

41. Isaiah Berlin, "My Intellectual Path," *The Power of Ideas* (Princeton, NJ: Princeton University Press, 2000), quoted in Graham et al., "Moral Foundations Theory," 4.

42. Graham et al., "Moral Foundations Theory," 5.

CHAPTER 6

1. Fiery Cushman, "Morality: Don't Be Afraid—Science Can Make Us Better," *New Scientist*, October 13, 2010, www.newscientist.com.

2. See, for instance, the response essays in the July 2014 (vol. 124, no. 4) issue of philosophy's flagship journal for moral theory, *Ethics*, and Erik Wielenberg's book review of Greene's *Moral Tribes* in the same issue.

3. Joshua D. Greene, R. Brian Sommerville, Leigh E. Nystrom, John M. Darley, and Jonathan D. Cohen, "An fMRI Investigation of Emotional Engagement in Moral Judgment," *Science* 293, no. 5537 (September 14, 2001): 2105–8, doi:10.1126/science.1062872; Joshua D. Greene, Leigh E. Nystrom, Andrew D. Engell, John M. Darley, and Jonathan D. Cohen, "The Neural Bases of Cognitive Conflict and Control in Moral Judgment," *Neuron* 44, no. 2 (October 14, 2004): 389–400, doi:10.1016/j.neuron.2004.09.027.

4. Recent studies allege among other things that people with prefrontal cortex damage tend to make more characteristically utilitarian judgments, and that drunk people tend to do so as well. See M. Koenigs, L. Young, R. Adolphs, D. Tranel, F. Cushman, M. Hauser, & A. Damasio, "Damage to the Prefrontal Cortex Increases Utilitarian Moral Judgments," *Nature* 446, no. 7138 (2007): 908–11, http://doi.org/10.1038/nature05631; Aaron A. Duke and Laurent Bègue, "The Drunk Utilitarian: Blood Alcohol Concentration Predicts Utilitarian Responses in Moral Dilemmas," *Cognition*, vol. 134 (2015): 121–27.

5. Joshua Greene, *Moral Tribes: Emotion, Reason, and the Gap Between Us and Them* (New York: Penguin Press, 2013), 329.

6. See Guy Kahane, "Intuitive and Counterintuitive Morality," in *Moral Psychology and Human Agency: Philosophical Essays on the Science of Ethics* (Oxford: Oxford University Press, 2014), 9–34.

7. Kahane, "Intuitive and Counterintuitive Morality," 14.

8. Guy Kahane, Katja Wiech, Nicholas Shackel, Miguel Farias, Julian Savulescu, and Irene Tracey, "The Neural Basis of Intuitive and Counterintuitive Moral Judgment," *Social Cognitive and Affective Neuroscience* 7, no. 4 (2012): 393–402.

9. The overreaching extends beyond what we reviewed in the body of this chapter. Even if we suppose that this dual-process theory of moral judgment is correct (and scientifically well-confirmed), what does Greene want to build from this foundation? What are the objectives? In fact, there are several. One is to make the case that science "can advance ethics by revealing the hidden inner workings of our moral judgments, especially the ones we make intuitively. Once those inner workings are revealed we may have less confidence in some of our judgments and the ethical theories that are (explicitly or implicitly) based on them" (Joshua D. Greene, "Beyond Point—and—Shoot

Morality: Why Cognitive (Neuro)Science Matters for Ethics," *Ethics* 124, no. 4 [2014]: 695–96). Another objective is to establish that scientific study of the brain yields nontrivial contributions to moral theory, traditionally the province of philosophers. Yet another, more ambitious objective is to chart a path for overcoming global moral disagreement by making a case for utilitarianism and against moralities that give an important place to rights and duties. Finally, perhaps most ambitiously, Greene hopes that the scientific study of the mind will—by revealing that it functions as a machine that is intelligible from a third-person perspective—decisively rule out substance dualism, the view that human beings have a soul in addition to a physical body. See Joshua D. Greene, "Social Neuroscience and the Soul's Last Stand," in *Social Neuroscience: Toward Understanding the Underpinnings of the Social Mind*, ed. A. Todorov, S. Fiske, and D. Prentice (New York: Oxford University Press, 2011). Greene's arguments both for how to overcome global moral disagreement and against substance dualism are almost wholly philosophical rather than empirical. Unsurprisingly, they are hugely controversial and are supported by scientific evidence only as much as are the opposing positions.

10. Frans B. M. de Waal, *The Age of Empathy: Nature's Lessons for a Kinder Society* (New York, Harmony, 2009).

11. Frans B. M. de Waal, *Primates and Philosophers: How Morality Evolved* (Princeton, NJ: Princeton University Press, 2009).

12. Steven Pinker, "The Moral Instinct," *New York Times*, January 13, 2008, www.nytimes.com.

13. Pinker, "Moral Instinct."

14. Leon R. Kass, "The Wisdom of Repugnance," *New Republic* 216, no 22 (June 2, 1997).

15. Of course, there is a great deal of scientific overreaching in popular discourse. Consider, for instance, Neil deGrasse Tyson's comments concerning his proposed virtual nation, Rationalia. In Rationalia—surmises Tyson—all policy decisions are to be based on evidence. This might sound good, but only until you dig a bit deeper into what that might look like. Tyson says, "In Rationalia, if you want to fund art in schools, you simply propose a reason why. Does it increase creativity in the citizenry? Is creativity good for culture and society at large? Is creativity good for everyone no matter your chosen profession? These are testable questions. They just require verifiable research to establish answers. And then, the debate ends quickly in the face of evidence, and we move on to other questions" ("Reflections on Rationalia," Facebook post, August 7, 2016, https://www.facebook.com/notes/neil-degrasse-tyson/reflections-on-rationalia/10154399608556613/). Tyson is light on the details for how this sort of testing might go, even in the broadest of strokes. In the

end, Tyson's thoughts on ethical experimentation never move beyond the level of superficial speculation.

16. Michael Kosfeld, Markus Heinrichs, Paul J. Zak, Urs Fischbacher, and Ernst Fehr, "Oxytocin Increases Trust in Humans," *Nature* 435 (June 2, 2005): 673–76, doi:10.1038/nature03701.

17. At the time of this writing, the citation count was over thirty-three hundred.

18. Paul J. Zak, "Trust, Morality, — and Oytocin?" July 2011, https://www.ted.com, over 1.6 million views at the time of this writing; book: *The Moral Molecule: The Source of Love and Prosperity* (New York: Dutton, 2012).

19. For an amusing discussion of Zak's questionable affectations in presenting his research, see the May 8, 2016, episode of HBO's *Last Week Tonight with John Oliver*.

20. Joyce Berg, John Dickhaut, and Kevin McCabe, "Trust, Reciprocity, and Social History," *Games and Economic Behavior* 10 (1995): 122–42, 123.

21. So, for instance, A might be initially allotted $10 and choose to give $3 to B; the gift is then tripled to $9, which, added to the $10 the B subject started with, gives him or her a total of $19, and finally B chooses to give $5 back to A. So A started with $10 and leaves with $12, and B started with $10 and leaves with $14. Both benefit.

22. Zak, *Moral Molecule*, 64.

23. Zak, *Moral Molecule*, 65–66.

24. Zak, *Moral Molecule*, xi.

25. Zak, *Moral Molecule*, xvi–xvii.

26. Gideon Nave, Colin Camerer, and Michael McCullough, "Does Oxytocin Increase Trust in Humans? A Critical Review of Research," *Perspectives on Psychological Science* 10, no. 6 (November 2015): 772–89.

27. Quoted in Ed Yong, "The Weak Science behind the Wrongly Named Moral Molecule," *The Atlantic*, November 13, 2015, www.theatlantic.com. Yong has published several pieces helpfully aggregating the damning scientific evidence now confronting Zak's project. In addition to the piece just cited, see Yong's "One Molecule for Love, Morality, and Prosperity?" *Slate*, July 17, 2012, www.slate.com.

28. Helen Shen, "Neuroscience: The Hard Science of Oxytocin," *Nature* 522 (June 25, 2015): 410–12, doi:10.1038/522410a.

29. Tamsin Shaw, Steven Pinker, and Jonathan Haidt, "Moral Psychology: An Exchange." *New York Review of Books*, April 7, 2016, http://www.nybooks.com.

30. Jonathan Haidt, *The Righteous Mind: Why Good People Are Divided by Politics and Religion*, reprint ed. (New York: Vintage, 2013), 271.

31. Steven Pinker, *The Better Angels of Our Nature: Why Violence Has Declined*, reprint ed. (New York Penguin Books, 2012), 623. Similarly, in his latest

work, he admits that "the scientific facts do not by themselves dictate values..." *Enlightenment Now: The Case for Reason, Science, Humanism, and Progress* (New York: Viking, 2018), 394

32. Greene, *Moral Tribes*, 189. However, Greene denies that there is any such thing as moral truth, so his denials here would seem to reflect this fact more than any caution about proclaiming moral results from scientific inquiry. See also Greene, *Moral Tribes*, 186.

33. Pinker, *Enlightenment Now*, 395.

34. Greene, *Moral Tribes*, 189. See also Greene, "Beyond Point-and-Shoot Morality," 711.

35. This is an observation made by Thomas Nagel as well in his review of *The Righteous Mind*. See Thomas Nagel, "The Taste for Being Moral," *New York Review of Books*, December 6, 2012.

36. See http://www.ethicalsystems.org/content/ethics-pays.

37. Other scholars do the same, most obviously those who consult for Ethicalsystems.org. Consider also Marc Hauser's postacademic career as the founder of the company Risk Eraser. In his own words, "As I turn to this second chapter in my career, my only hope is that my skills can help enrich the programs that help the at-risk population, so that these deserving children grow to make wise choices and lead healthier and more meaningful lives. As the company name reveals, I want to help erase the risk in at-risk children" (http://www.risk-eraser.com/ourteam).

38. Once again, see Fiery Cushman, "Morality: Don't Be Afraid—Science Can Make Us Better," *New Scientist*, October 13, 2010, https://www.newscientist .com.

39. See Drake Bennett, "Ewwwwwwwww! The Surprising Moral Force of Disgust," *Boston Globe*, August 15, 2010, http://archive.boston.com. See also https://www.edge.org/event/the-new-science-of-morality.

40. "This book will help you flourish. There, I have finally said it. I have spent my professional life avoiding unguarded promises like this one. I am a research scientist, and a conservative one at that. The appeal of what I write comes from the fact that it is grounded in careful science: statistical tests, validated questionnaires, thoroughly researched exercises, and large, representative samples. In contrast to pop psychology and the bulk of self-improvement, my writings are believable because of the underlying science" (Martin Seligman, *Flourish: A Visionary New Understanding of Happiness and Well-Being* [New York: Free Press, 2011], 1).

For a few more illustrations of this, see Daniel Nettle, *Happiness: The Science behind Your Smile* (New York: Oxford University Press, 2006); Gabriele Oettingen, *Rethinking Positive Thinking: Inside the New Science of Motivation* (New York: Penguin, 2015); Loretta Graziano Breuning, *The Science of Positivity*

(Avon, MA: Adams Media, 2017); Neil Neimark, *The Science of Positive Thinking* (Irvine, CA: Author, 2015); Emma Seppala, *The Happiness Track: How to Apply the Science of Happiness to Accelerate Your Success* (New York: HarperOne, 2016).

CHAPTER 7

1. Among *many* others, see, for instance, Christopher Boehm, *Moral Origins: The Evolution of Virtue, Altruism, and Shame* (New York: Basic Books, 2012), and Robert Wright, *The Moral Animal: The New Science of Evolutionary Psychology* (New York: Vintage Books, 1994).

2. Boehm's descriptive and merely biological definition of morality—in terms of altruis—is buried in *Moral Origins*, 367n31.

3. Patricia Churchland, *Braintrust: What Neuroscience Tells Us about Morality* (Princeton, NJ: Princeton University Press, 2011), 2–3.

4. Churchland, *Braintrust*, 3. As she puts it, "Morality can be—and I argue, is—grounded in our biology, in our capacity for compassion and our ability to learn and figure things out" (Churchland, *Braintrust*, 200).

5. Churchland, *Braintrust*, 3.

6. Churchland, *Braintrust*, 4.

7. Churchland, *Braintrust*, 8–9.

8. Churchland, *Braintrust*, 200.

9. Churchland, *Braintrust*, 59.

10. Churchland, *Braintrust*, 9–10, 59. In some places, it's unclear whether she ultimately thinks there is any identifiable line demarcating the moral from the merely social: "That nonhuman mammals have social values is obvious; they care for juveniles, and sometimes mates, kin, and affiliates; they cooperate, they may punish, and they reconcile after conflict. We could engage in a semantic wrangle about whether these values are really moral values, but a wrangle about words is apt to be unrewarding. Of course only humans have human morality. But that is not news, simply a tedious tautology. One might as well note that only marmosets have marmosets morality, and so on down the line" (Churchland, *Braintrust*, 26). Her thought here appears to be that what separates the moral social behaviors of humans from those of similar mammals is just what species happens to possess them.

11. David Sloan Wilson, *Does Altruism Exist?* (New Haven, CT: Yale University Press, 2015), 3.

12. Our view is roughly in line with Samir Okasha's in his essay on the subject in *The Stanford Encyclopedia of Philosophy*. He writes,

> Ordinarily we think of altruistic actions as disinterested, done with the interests of the recipient, rather than our own interests, in mind. But kin selection theory explains altruistic behaviour as a clever strat-

egy devised by selfish genes as a way of increasing their representation in the gene-pool, at the expense of other genes. . . . The key point to remember is that biological altruism cannot be equated with altruism in the everyday vernacular sense. Biological altruism is defined in terms of fitness consequences, not motivating intentions. If by "real" altruism we mean altruism done with the conscious intention to help, then the vast majority of living creatures are not capable of "real" altruism nor therefore of "real" selfishness either. ("Biological Altruism," *Stanford Encyclopedia of Philosophy* [Fall 2013 edition], ed. Edward N. Zalta, http://plato.stanford.edu)

13. Wilson, *Does Altruism Exist?* 141.

14. Wilson, *Does Altruism Exist?* 142.

15. Additionally, as some philosophers have pointed out, virtue theory need not claim that people *are* or *tend to be* virtuous, acting from stable character traits. Rather, all the virtue theorist is committed to is that such traits are what goodness for a person *consists in*—it may well turn out that there are actually few good people, few people who have virtuous character traits (N. Athanassoulis, "A Response to Harman: Virtue Ethics and Character Traits," *Proceedings of the Aristotelian Society* 100 [2000]: 215–22; J. J. Kupperman, "The Indispensability of Character," *Philosophy* 76 [2001]: 239–50; M. DePaul, "Character Traits, Virtues, and Vices: Are There None?" *Proceedings of the 20th World Congress of Philosophy*, vol. 1 [Bowling Green, OH: Philosophy Documentation Center, 1999]).

16. Gopal Sreenivasan, "Errors about Errors: Virtue Theory and Trait Attribution," *Mind* 111 (January 2002): 47–68.

17. Hugh Hartshorne and Mark A. May, *Studies in the Nature of Character*, vol. 1, *Studies in Deceit* (New York: Macmillan, 1928).

18. Sreenivasan, "Errors about Errors," 60.

19. Grotius, *The Rights of War and Peace*, Preliminary Discourse, 40, 110–11.

20. Joshua Greene, *Moral Tribes: Emotion, Reason, and the Gap between Us and Them* (New York: Penguin Press, 2013), 302, 304, 305.

21. Michael Shermer, *The Moral Arc: How Science Leads Humanity toward Truth, Justice, and Freedom* (New York: Henry Holt, 2015).

22. "Can Science Determine Moral Values? A Challenge from and Dialogue with Marc Hauser about The Moral Arc," on Shermer's blog "The Moral Arc," http://moralarc.org.

23. Sam Harris, *The Moral Landscape: How Science Can Determine Human Values* (New York: Free Press, 2010), 37.

24. Harris's rhetorical legerdemain on this point can be seen in an exchange with a graduate student during a Q&A session after a presentation at Oxford University. The student insightfully points out that while Harris

has promised scientific demonstration of moral values, he instead merely appeals to common sense to support his fundamental value claim about well-being. Harris's response begins, "The moment you grant that . . . we're right to talk about well-being . . . , then all of the facts that determine well-being become the facts of science." But this is precisely the move the student is objecting to. *Why, scientifically,* should we grant that "we're right to talk about well-being"? After this opening, dodge Harris grandstands for another four minutes on the ways science might get at the facts of well-being once we've assumed its value, but he never responds to the student's question. All the work of establishing the value of well-being is shoehorned into his opening phrase, "The moment you grant"—a subtle shift from demonstration to common sense that few will catch. Unfortunately, the student did not appear to get a follow-up question to call out Harris for his evasion. See the exchange here: https://www.youtube.com /watch?v=UuuTOpZxwRk&t=.

25. Martin Seligman, *Flourish: A Visionary New Understanding of Happiness and Well-Being* (New York: Free Press, 2011), 10.

26. Seligman, *Flourish*, 16–20. Seligman prefers to use the term "happiness" for subjective positive emotion, but his broader account acknowledges a richer view traditionally considered to be a candidate for what happiness is. He notes this on page 11: "Happiness historically is not closely tied to such hedonics."

27. See Joseph Henrich, Steven Heine, and Ara Norenzayan, "Most People Are Not WEIRD," *Nature* 466, no. 1 (2010): 29; Joseph Henrich, Steven Heine, and Ara Norenzayan, "The Weirdest People in the World," *Behavioral and Brain Sciences* 33, nos. 2–3 (June 2010): 61–83; Jeffrey Arnett, "The Neglected 95%: Why American Psychology Needs to Become Less American," *American Psychologist* 63, no. 7: 602–14. The problem is even worse (or weirder) than this because the subjects are overwhelmingly drawn from undergraduate student populations.

28. The question favored by Kahneman and others is "Taking all things together, how satisfied are you with *your life as a whole* these days?" See Daniel Kahneman and Alan B. Krueger, "Developments in the Measurement of Subjective Well-Being," *Journal of Economic Perspectives* 20, no. 1 (Winter 2006): 3–24 (emphasis in the original).

29. This helps to explain why happiness scholarship has turned to self-reporting. But the problem here, among other things, is that it is tough to tell whether the same concept is being identified as "happiness" across different subjects.

30. Pawelski here is summarizing an argument made by Darrin McMahon in "The Pursuit of Happiness in History," in *The Oxford Handbook of Happiness*, ed. Susan A. David, Ilona Boniwell, and Amanda Conley Ayers (Oxford:

Oxford University Press, 2013). Pawelski's comment comes from the same volume, "Introduction to Philosophical Approaches to Happiness" (248).

31. Martha Nussbaum, "Who Is the Happy Warrior? Philosophy Poses Questions to Psychology," *Journal of Legal Studies*, 37, S2 (June 2008): S93.

32. One (veiled) attempt comes from the social psychologists Fiery Cushman and Liane Young. In their paper "Patterns of Moral Judgment Derive from Nonmoral Psychological Representations," *Cognitive Science* 35 (2011), Cushman and Young attempt to use statistical analysis of questionnaire results to determine what role moral principles play in people's moral reasoning. Cushman and Young claim to show that people often classify their moral judgments as instances of moral principles but often employ nonmoral distinctions in making these judgments.

As Cushman and Young put it,

> The distinctions between action versus omission, means versus side-effect and contact versus noncontact affect moral judgments indirectly, by way of psychological representations regularly deployed in nonmoral domains of cognition.... Moral principles (e.g., action versus omission, means versus side-effect) are thus derived from nonmoral psychological representations. This "derived" approach is congruent with a basic ambition of attribution theory: to explain moral judgment in terms of antecedent, nonmoral assessments of causal responsibility and intent. ("Patterns of Moral Judgement," 1054)

From our vantage point, their most interesting claim concerns how we draw moral conclusions. But even here, it isn't clear what Cushman and Young claim to be showing. On the one hand, it sounds like their claim is that some moral judgments are sensitive to certain nonmoral judgments. If that's their claim, it doesn't require scientific confirmation—who would doubt it? Moral judgments are often sensitive to nonmoral judgments. For instance, take any judgment that theft has occurred. This is of course a moral claim. But judging that theft has taken place also requires a nonmoral judgment that something has changed hands. After all, if everybody still has what they had before, no theft has occurred. If this is all Cushman and Young mean to say here, then this is thoroughly unremarkable.

On the other hand, when Cushman and Young say that "moral principles ... are thus derived from nonmoral psychological representations," it sounds as if they have demonstrated that moral principles can be derived from nonmoral phenomena—that some "oughts" really can be derived from an "is."

But they have shown no such thing. None of their experiments show any way of getting from nonmoral judgments of, say, causal responsibility to moral judgments without reliance upon specifically moral judgments or ideas ("psychological representations"). Yes, some moral judgments do rely *in part* on nonmoral judgments or ideas—recall the theft example above. But this doesn't mean we can derive moral judgments solely from nonmoral judgments. For all their experiments show, a moral judgment or psychological representation is required *along with* a nonmoral judgment in order to generate a moral judgment of the situation. In other words, they haven't shown that we can get to a moral conclusion—an "ought"—just from nonmoral thoughts or concepts— from an "is."

The upshot, then, is that their claim about the relationship of moral judgments to nonmoral judgments is either trivial or unsupported. Their case appears successful only because it trades on ambiguous descriptions about what is being shown.

CHAPTER 8

1. Owen Flanagan, *The Really Hard Problem* (Cambridge, MA: MIT Press, 2007), 107.

2. Philip Kitcher, *The Ethical Project* (Cambridge, MA: Harvard University Press, 2011), 2. He also sees "the ethical project as begun by our remote ancestors, in response to the difficulties of their social life. They invented ethics."

3. Fiery Cushman, "Morality: Don't Be Afraid—Science Can Make Us Better," *New Scientist*, October 13, 2010, www.newscientist.com. Cushman, along with Joshua D. Greene, add elsewhere, "Just as theories in scientific domains will tend to reflect the structure of the world, there is reason to suppose that moral philosophies will tend to reflect the structure of the mind" ("The Philosopher in the Theater," in *The Social Psychology of Morality: Exploring the Causes of Good and Evil*, ed. M. Mikulincer & P. R. Shaver [Washington, DC: American Psychological Association, 2012].

   Some might wonder: aren't all of our concepts man-made products of convention? Not in this sense. Sure, human language–users decide which symbols and which sounds stand for which things, but the sort of conventionality at stake here is different. Consider, for instance, water. It's conventional that English-speakers call water "water," and that Spanish-speakers call it "agua." But the entity being referred to—H_2O—is a real substance and has the chemical properties it does regardless of what anyone calls it. If we had decided to group alcohol and water together under the same name, there would nevertheless be a real difference between the two substances. Water is a real thing, with its own distinctive features, no matter what anyone thinks or calls it.

   For the philosophical naturalists under discussion here, morality is *not* like this. For them, morality is more like etiquette: if a given culture went out of existence, its rules of etiquette would disappear as well. And if a given culture decided to change the rules of etiquette, they could do so, and the truths about etiquette would change along with their conventions.

4. "Normativity for Naturalists," *Philosophical Issues* 25, no. 1.

5. Kitcher, *Ethical Project*, 263.

6. Patricia Churchland, *Braintrust: What Neuroscience Tells Us about Morality* (Princeton, NJ: Princeton University Press, 2011), 12–16.

7. Michael S. Gazzaniga, *The Ethical Brain: The Science of Our Moral Dilemmas* (New York: Harper-Perennial, 2006), 171–72.

8. See Joshua Greene, *Moral Tribes: Emotion, Reason, and the Gap between Us and Them* (New York: Penguin Press, 2013), 172, as well as Philip Kitcher, who says:

> There is no ethical justification for following a rule unless one has grounds for viewing that rule as authoritative, and those grounds can come not from labeling the source—either as divine lawgiver or as its detheologized counterpart "the moral law within"—but only from recognizing the rule as well adapted to producing good outcomes. Following rules not well adapted to producing good outcomes is a capricious and irresponsible thing to do: that is why consequentialism is "the doctrine of rational persons of all schools." Ungrounded deontology is dangerous. (Kitcher, *Ethical Project*, 289)

The bit Kitcher quotes here is from J. S. Mill.

9. Churchland, *Braintrust*, 190.

10. Paul Thagard, *The Brain and the Meaning of Life* (Princeton, NJ: Princeton University Press, 2010), 184–88.

11. See Max Weber, "Science as a Vocation," *From Max Weber: Essays in Sociology*, trans. and ed. H. H. Gerth and C. Wright Mills (New York: Oxford University Press, 1946). Alex Rosenberg also identifies his view under this name. See, for instance, his "Disenchanted Naturalism," in *Contemporary Philosophical Naturalism and Its Implications*, ed. Bana Bashour and Hans Muller (New York: Routledge, 2014).

12. Thagard, *Brain and the Meaning of Life*, 8–9.

13. See, for instance, Dean Zimmerman's "The Privileged Present: Defending an 'A-Theory' of Time," in *Contemporary Debates in Metaphysics*, ed. Ted Sider, John Hawthorne, and Dean Zimmerman (Malden, MA: Blackwell, 2008), 217; Theodore Sider, *Four Dimensionalism: An Ontology of Persistence and Time* (New York: Oxford University Press, 2001), 41–42; or J. L. Mackie's famous judgment that morality is "queer" (*Ethics: Inventing Right and Wrong* [London: Pelican Books, 1977]).

14. See Alva Noë's animadversions on the spooky conception of life in his review of Nagel, "Are the Mind and Life Natural?" October 12, 2012, www.npr.org.

15. Jerry Fodor, *A Theory of Content and Other Essays* (Cambridge, MA: Bradford Books / MIT Press, 1990), 156.

16. Daniel Dennett, "The Self as a Center of Narrative Gravity," in *Self and Consciousness: Multiple Perspectives*, ed. F. Kessel, P. Cole, and D. Johnson (Hillsdale, NJ: Erlbaum, 1992); Gary Gutting, "Sam Harris's Vanishing Self," New York Times, September 7, 2014, https://opinionator.blogs.nytimes.com.

17. Daniel Dennett, *From Bacteria to Bach and Back: The Evolution of Minds* (New York: Norton, 2017); Alex Rosenberg, *The Atheist's Guide to Reality: Enjoying Life without Illusions* (New York: Norton, 2011).

18. Disenchanted naturalists sometimes speak of various features of this or that organism as having been selected *for* its adaptive function. But this is metaphorical; they don't mean that there's any genuine teleology present in evolution.

19. Here Leiter is explaining a point about naturalism made by Thomas Scanlon (not himself a naturalist). But the point is one that Leiter is endorsing. See "Normativity for Naturalists."

20. Joshua D. Greene, "The Terrible, Horrible, No-Good, Very Bad Truth about Morality and What to Do about It," PhD diss. (Princeton: Princeton University, 2002), 54–141.

21. In his book *The Moral Landscape: How Science Can Determine Human Values* (New York: Free Press, 2010), Harris says that he is not merely claiming that science can tell us what people happen to think morality is, nor is he merely arguing that science can help us reach our goals. Rather, he says that "science can, in principle, help us understand what we *should* do and *should* want—and, therefore, what *other people* should do and should want in order to live the best lives possible" (28). His argument goes like this: The only thing that could ultimately have any value is the well-being of conscious beings. So, all moral truths are ultimately about well-being. But these truths about well-being are grounded, "in some lawful and not entirely arbitrary way, to states of the human brain and to states of the world" (15). And the brain and world can be studied scientifically, "regarding positive and negative social emotions, retributive impulses, the effects of specific laws and social institutions on human relationships, the neurophysiology of happiness and suffering, etc." (1–2). "Once we see that a concern for well-being . . . is the only intelligible basis for morality and values, we will see that there must be a science of morality, whether or not we ever succeed in developing it: because the well-being of conscious creatures depends upon how the universe is, altogether. Given that changes in the physical universe and in our experience of it can be understood, science should increasingly enable us to answer specific moral questions" (28). He recognizes that this commits him "to some form of moral realism (viz. moral claims can really be true or false)" (62; see also 8).

22. Harris acknowledges his own marginality on this point: "Most educated, secular people (and this includes most scientists, academics, and journalists) believe that there is no such thing as moral truth—only moral preference, moral opinion, and emotional reactions that we mistake for genuine knowledge of right and wrong" (Moral *Landscape*, 29; see also 27).

23. Owen Flanagan, comments at the Moving Naturalism Forward conference, October 27, 2012, morning session of day two, http://preposterousuniverse. com/naturalism2012/video.html, roughly at the 1:01:00 mark. Flanagan elaborates his views in his *Really Hard Problem*, 125. In a similar vein, Patricia Churchland writes, ". . . what distinguishes moral values from other values. I generally shy away from trying to cobble together a precise definition of 'moral,' preferring to acknowledge that there is a spectrum of social behaviors, some of which involve matters of great seriousness, and tend to be called moral, such as enslaving captured prisoners or neglecting children, while others involve matters or more minor moments, such as conventions for behavior at a wedding" (*Braintrust*, 9–10). The important point here, for our purposes, is that Churchland situates the moral as a species of the social. What separates the moral from the merely social? What we call things and what we take more seriously. Similarly, she says, "Social behavior and moral behavior appear to be part of the same spectrum of actions, where those actions we consider 'moral' involve more serious outcomes than do merely social actions such as bringing a gift to a new mother" (*Braintrust*, 59).

24. Not all naturalists subsume the language of moral oughts into the nonmoral oughts of prudence or practicality. Some—including Rosenberg—think that no oughts whatsoever, whether moral or nonmoral, are real features of the world, and that moral language must be understood in some other way. See Leiter "Normativity for Naturalists," for a brief description of alternatives and references, and Rosenberg's *Atheist's Guide to Reality*, 94–145.

25. Similar results would appear to follow for ethical properties more generally.

26. "It seems the naturalist has a straightforward account of normativity: what we call normativity is simply an artifact of the psychological properties of certain biological organisms, i.e., what they feel or believe or desire (or are disposed to feel, believe, or desire). As long as the posited organisms are naturalistically respectable, and the mental states invoked are as well, then that is the end of the naturalist's story" (Leiter, "Normativity for Naturalists").

27. Richard Tuck, "Grotius, Carneades, and Hobbes," *Grotiana* 4 (new series) (1983): 43–62; Jerome Schneewind, *The Invention of Autonomy* (Cambridge: Cambridge University Press, 1998), 82.

28. Henry Sidgwick, *Outlines of a History of Ethics* [5th ed., 1902; repr. Indianapolis, IN: Hackett Publishing Company, 1988], 223.

29. Sidgwick, *Outlines*, 224.

30. Joshua Greene, "The Terrible, Horrible, No-Good, Very Bad Truth about Morality and What to Do About It," written under the supervision of ethical theorist Gilbert Harman and famed metaphysician David Lewis.
31. Greene, "Terrible, Horrible," 139–40.
32. Greene's argument for moral nihilism is essentially that given naturalism—that morality can be explained in terms of scientifically amendable, value-neutral properties—there's no metaphysically kosher way to connect up moral properties with moral rules. So his nihilistic arguments presuppose naturalism ("Terrible, Horrible," 139–40, 134–35).
33. "So is there moral truth? In my dissertation . . . I argued that there isn't any moral truth. . . . But now I think what matters most for practical purposes is the possibility of objective improvement, not the possibility of objective correctness. And this inclines me to say that, for practical purposes, there can be something very much like moral truth . . . which might more or less be the moral truth, if not the Moral Truth. But, really, I think it's the wrong question on which to focus, which is why I've paid it relatively little attention in this book. What matters is what we do with the morass [of moral disagreement], not whether we call the final product the 'moral truth'" (Greene, *Moral Tribes*, 373–74)
34. Rosenberg, *Atheist's Guide to Reality*, 94–96.
35. Rosenberg, *Atheist's Guide to Reality*, 102.
36. Rosenberg, *Atheist's Guide to Reality*, 109.
37. Rosenberg supports this claim in part by saying,

> The physical facts fix all the facts, including the biological ones. These in turn have to fix the human facts—the facts about us, our psychology, and our morality. After all, we are biological creatures, the result of a biological process that Darwin discovered but that the physical facts ordained. As we have just seen, the biological facts can't guarantee that our core morality (or any other one, for that matter) is the right one, true, or correct one. If the biological facts can't do it, then nothing can. No moral core is right, correct, true. That's nihilism. And we have to accept it. (*Atheists Guide to Reality*, 113)

38. Rosenberg, *Atheist's Guide to Reality*, 94–96.
39. Philip Kitcher, "Naturalistic Ethics without Fallacies," *Preludes to Pragmatism: Toward a Reconstruction of Philosophy* (New York: Oxford University Press, 2012), 304.
40. Flanagan, *Really Hard Problem*, 2; also, "Naturalism is impressed by the causal explanatory power of the sciences. Science typically denies the truth—or at least the testability—of theories that invoke non-natural, occult, or supernatural causes or forces" (Flanagan, *Really Hard Problem*, 2).

41. He continues:

> Here is the possibility proof:
> 1. Humans are natural creatures who live in the natural world.
> 2. According to the neo-Darwinian consensus, humans are animals: *Homo sapiens sapiens*, mammals who know and know that they know.
> 3. Human practices are natural phenomena.
> 4. Art, science, ethics, religion, and politics are human practices.
> 5. The natural sciences and the human sciences can, in principle, describe and explain human nature and human practices.
> 6. Therefore, the sciences can explain, in principle, the nature and the function of art, science, ethics, religion, and politics. (Flanagan, *Really Hard Problem*, 21)

Additional commentary from Flanagan clarifying that he is committed to scientific reductionism:

> What might explaining [art, science, ethics, religion, and politics]—our ways of worldmaking—involve? Presumably we would try to understand the nature and functions of these practices, as well as their causal antecedents and consequences. This would lead us to understand the nature of *Homo sapiens* more deeply. It would almost inevitably require changes in traditional narratives of self-understanding.... One surprisingly common idea is that science, in explaining some phenomenon, makes it something it isn't or wasn't. It tries to disclose that everything is a "mere thing." It takes the world as we know it and turns it into a mere collection of scientific objects. "Reductionism" is the disparaging name for this phenomenon. Something like this view—that reduction always entails that things are not as they seem, and that such phenomena as consciousness are revealed as illusory—is common. But it rests on a mistake. To say that some phenomenon can be understood scientifically, even that it can be reduced, is not to say that the phenomenon is itself "scientific," nor does it entail that the phenomenon we began with disappears or evaporates—whatever exactly that might mean—when we get at its deep structure. (Flanagan, *Really Hard Problem*, 21–22)

42. Flanagan, *Really Hard Problem*, 23.
43. Flanagan, "Varieties of Naturalism," in *The Oxford Handbook of Religion and Science*, ed. Philip Clayton and Zachary Simpson (Oxford: Oxford University Press, 2006), 446n19.
44. Patricia Churchland, "Moral Decision-Making and the Brain," *Oxford Handbook of Neuroethics*, ed. J. Illes and B. J. Sahakian (New York: Oxford, 2011), 3.
45. Someone might object that some actions will still be better or worse with

respect to what someone is trying to accomplish, or with respect to adherence to conventional rules. This is true, but the sense of "better" and "worse" here carry no moral weight—they have no normative force. After all, plans can be better and worse given the goal of robbing a bank. The point is that "better" and "worse" understood in this way are only really good or bad insofar as the goal in question is good or bad. But good and bad goals—as real, objective states of affairs—are exactly the sorts of things that don't exist given this moral naturalism.

46. Greene, *Moral Tribes*, 290.
47. Cushman, "Don't Be Afraid."
48. Greene, "Terrible, Horrible."
49. Flanagan, *Really Hard Problem*, 121.
50. Flanagan, comments at the Moving Naturalism Forward conference. Flanagan elaborates on his views in *Really Hard Problem*, 125.
51. Flanagan, *Really Hard Problem*, 1–3.
52. Alex Rosenberg, "Disenchanted Naturalism," *Kritikos* 12 (January–April 2015), emphasis added.
53. Greene, *Moral Tribes*, 189.
54. Greene, *Moral Tribes*, 191.
55. Greene, *Moral Tribes*, 291.
56. Greene, *Moral Tribes*, 291.
57. Flanagan, *Really Hard Problem*, 38–39, emphasis added.
58. Greene, *Moral Tribes*, 291. "I believe that the values behind utilitarianism are our true common ground.... We ... are united by our capacity for positive and negative *experience*, for happiness and suffering, and by our recognition that morality must, at the highest level, be *impartial*" (*Moral Tribes*, 189). "Happiness is what matters, and everyone's happiness counts the same," and so "We should simply try to make the world as happy as possible" (*Moral Tribes*, 170, 333).
59. Greene, *Moral Tribes*, 189.
60. Greene, *Moral Tribes*, 156–161.
61. "Put them together and our task, insofar as we're moral, is to make the world as happy as possible, giving equal weight to everyone's happiness. I do not claim, however, that utilitarianism is the moral truth. Nor do I claim, more specifically, and as some readers might expect me to, that science proves that utilitarianism is the moral truth. Instead, I claim that utilitarianism becomes uniquely attractive once our moral thinking has been *objectively improved* by a scientific understanding of morality" (Greene, *Moral Tribes*, 189).
62. In certain ways, this position echoes Rosenberg's inane contention that "Darwinian processes operating on our forebears in the main selected for niceness." See Rosenberg, "Disenchanted Naturalism." This, he says, "was

a convenience, not for us as individuals, but for our genes." He goes on to say that "after enough cycles the result is a nice bell-shaped distribution of niceness, with a small number of people at the extreme ends of unconditional altruism and egoistic sociopathy. It can't be helped, of course. Variation is the rule and there is really no way to stamp out the sociopathy. All we can do is protect ourselves from it."

63. This is the heart of Martin Seligman's book *Flourish: A Visionary New Understanding of Happiness and Well-Being* (New York: Free Press, 2011), where he argues that human flourishing consists of five components: positive emotion, engagement, relationships, a sense of meaningfulness, and accomplishment.

64. Flanagan, *Really Hard Problem*, 1–2, 3.

CHAPTER 9

1. As Michael Shermer put it, "Just because we cannot yet think of how science might resolve this or that moral conflict does not mean that the problem is an insoluble one. Science is the art of the soluble, and we should apply it where we can." Quoted in Michael Shermer, "The Science of Right and Wrong: Can Data Determine Moral Values?" *Scientific American*, January 1, 2011, www .scientificamerican.com.

2. John Stuart Mill, "Bentham," *Utilitarianism and On Liberty*, ed. M. Warnock, (Oxford, 2003), 61.

3. Jennifer Schuessler, "An Author Attracts Unlikely Allies," *New York Times*, February 7, 2013.

4. Schuessler, "Author Attracts Unlikely Allies."

5. Robert Paul Wolff, "What I Have Been Reading," The Philosopher's Stone, September 1, 2013, blog post, http://robertpaulwolff.blogspot.com.

6. Simon Blackburn, "Thomas Nagel: A Philosopher Who Confesses to Finding Things Bewildering," *New Statesman*, November 8, 2012.

7. Alva Noë, a philosopher at the University of California, Berkeley, said as much: "[Nagel] is questioning a certain kind of orthodoxy, and they are responding in the way the orthodox respond," quoted in Schuessler, "Author Attracts Unlikely Allies."

8. Max Weber, "Science as a Vocation," published as "Wissenschaft als Beruf," *Gesammlte Aufsaetze zur Wissenschaftslehre* (Tubingen: J. C. B. Mohr, 1922), 524–55. Originally a speech at Munich University, 1918, published in 1919 by Duncker & Humboldt (Munich). From H. H. Gerth and C. Wright Mills, trans. and ed. *From Max Weber: Essays in Sociology* (New York: Oxford University Press, 1946), 129–56.

9. Indeed, that social consensus generated a youth culture that had a reputation for ferocity and fanaticism. As Hitler put it in 1933, "My program for

educating youth is hard. Weakness must be hammered away. In my castles of the Teutonic Order a youth will grow up before which the world will tremble. I want a brutal, domineering, fearless, cruel youth. Youth must be all that. It must bear pain. There must be nothing weak and gentle about it. The free, splendid beast of prey must once again flash from its eyes. . . . That is how I will eradicate thousands of years of human domestication. . . . That is how I will create the New Order" (see http://www.historyplace.com/worldwar2/hitleryouth/). Hitler got his wish.

10. See Jackson Lears, "Get Happy," *The Nation*, November 6, 2013.

11. It appears that a wide range of resources are needed for such moral heroism. See, for example, Samuel P. Oliner and Pearl M. Oliner, *The Altruistic Personality: Rescuers of Jews in Nazi Europe* (New York: Free Press, 1988), as well as the work of William Damon, *The Power of Ideals* (New York: Oxford University Press, 2015), and (with Anne Colby) *Some Do Care* (New York: Free Press, 1994).

12. See chapter 3.

Bibliography

Anscombe, G. E. M. "Modern Moral Philosophy." *Philosophy* 33, no. 124 (January 1958).

Appiah, Kwame Anthony. *Experiments in Ethics* (Cambridge, MA: Harvard University Press, 2008).

Arnett, Jeffrey. "The Neglected 95%: Why American Psychology Needs to Become Less American." *American Psychologist* 63, no. 7 (October 2008).

Athanassoulis, Nafsika. "A Response to Harman: Virtue Ethics and Character Traits." *Proceedings of the Aristotelian Society* 100 (2000).

Bacon, Francis. *The Advancement of Learning*, ed. Joseph Devy (1605; New York: P. F. Collier and Son, 1901).

Bakhtin, Mikhail. "The Problem of Speech Genre." In *Speech Genre and Other Late Essays.* (Austin: University of Texas Press, 1986).

Barbeyrac, Jean. *An Historical and Critical Account of the Science of Morality, and the Progress It Has Made in the World, from the Earliest Times Down to the Publication of Pufendorf of the Law of Nature and Nations* (London: Printed for J. Walthoe, R. Wilkin, J and J. Bonwicke, S. Birt, T. Ward, and T. Osborn, 1729).

Bennett, Drake. "Ewwwwwwwww! The Surprising Moral Force of Disgust." *Boston Globe*, August 15, 2010. http://archive.boston.com.

Bentham, Jeremy. *Deontology Together with a Table of the Springs of Action and the Article on Utilitarianism*, ed. J. H. Burns and H. L. A. Hart (London: Athlone Press, 1977).

Bentham, Jeremy. *An Introduction to the Principles of Morals and Legislation* (Oxford: Clarendon Press, 1907).

Bentham, Jeremy. *Of Laws in General*, ed. H. L. A. Hart (London: Athlone Press, 1970).

Bentham, Jeremy. *The Rationale of Reward* (London: Robert Heward, 1830).

Bentham, Jeremy. *The Works of Jeremy Bentham*, 11 vols. (Edinburgh: William Tait, 1838–1843). http://oll.libertyfund.org/titles/2009#Bentham_0872-01_1090.

Bentham, Jeremy, and F. C. Montague. *A Fragment on Government* (Oxford: Oxford University Press, 1891).

Berg, Joyce, John Dickhaut, and Kevin McCabe. "Trust, Reciprocity, and Social History." *Games and Economic Behavior* 10 (1995).

Berlin, Isaiah. "My Intellectual Path." *The Power of Ideas* (Princeton, NJ: Princeton University Press, 2000).

Bernal, J. D. *Science in History* (London: C. A. Watts & Co., 1957).

Blackburn, Simon. "Thomas Nagel: A Philosopher Who Confesses to Finding Things Bewildering." *New Statesman*, November 8, 2012.

Bloch, Maurice. "Westermarck's Theory of Morality in His and Our Time." In *Westermarck*, Occasional Paper No. 44 of the Royal Anthropological Institute, ed. David Shankland (Herefordshire, England: Sean Kingston Publishing, 2014).

Boehm, Christopher. *Moral Origins: The Evolution of Virtue, Altruism, and Shame* (New York: Basic Books, 2012).

Brandhorst, Mario. "Naturalism and the Genealogy of Moral Institutions." *Journal of Nietzsche Studies* 40, no. 1 (Autumn 2010).

Breuning, Loretta Graziano. *The Science of Positivity.* (Avon, MA: Adams Media, 2017).

Brosnan, Sarah F., and Frans B. M. de Waal. "Fairness in Animals: Where to from Here?" *Social Justice Research* 25, no. 3 (September 5, 2012).

Brosnan, Sarah F., and Frans B. M. de Waal. "Monkeys Reject Equal Pay." *Nature* 425 (September 18, 2003).

Brownowski, Jacob, and Bruce Mazlish. *The Western Intellectual Tradition* (New York: Dorset Press, 1960).

Burnham, James. *The Managerial Revolution* (New York: John Day, 1941).

Burton, John Hill. *Life and Correspondence of David Hume* (Edinburgh: William Tait, 1846).

Burtt, E. A. *The Metaphysical Foundations of Modern Science*, rev. ed. (New York: Doubleday Anchor Books, 1954).

Carlin, Laurence. *The Empiricists: A Guide for the Perplexed* (London: Continuum, 2009).

Carlyle, Thomas. *On Heroes, Hero-Worship, and the Heroic in History*, ed. M. K. Goldberg, J. J. Brattin, and M. Engel (Berkeley: University of California Press, 1993).

Churchland, Patricia. *Braintrust: What Neuroscience Tells Us about Morality* (Princeton, NJ: Princeton University Press, 2011).

Churchland, Patricia. "Moral Decision-Making and the Brain." *Oxford Handbook of Neuroethics*, ed. J. Illes and B. J. Sahakian (New York: Oxford, 2011).

Condorcet, Marquis de (Nicolas de Caritat). *Sketch for a Historical Picture of the Progress of the Human Mind*, trans. June Barraclough (London: Weidenfeld and Nicolson [University Microfilms], 1955).

Confer, Jaime C., Judith A. Easton, Diana S. Fleischman, Cari D. Goetz, David M. G. Lewis, Carin Perilloux, David M. Buss. "Evolutionary Psychology: Controversies, Questions, Prospects, and Limitations." *American Psychologist* 65, no. 2 (2010).

Coyne, Jerry. "New Paper Shows That Nowak et al. Were Wrong: Kin Selection Remains a Valuable Concept in Evolutionary Biology." *Why Evolution Is True*, March 27, 2015, https://whyevolutionistrue.wordpress.com.

Crimmins, James. *Secular Utilitarianism: Social Science and the Critique of Religion in the Thought of Jeremy Bentham* (New York: Oxford University Press, 1990).

Cushman, Fiery. "Morality: Don't Be Afraid—Science Can Make Us Better." *New Scientist* no. 2782 (October 13, 2010).

Cushman, Fiery, and Liane Young. "Patterns of Moral Judgment Derive from Non-moral Psychological Representations." *Cognitive Science* 35 (2011).

Damon, William. *The Power of Ideals* (New York: Oxford University Press, 2015).

Damon, William, and Anne Colby. *Some Do Care* (New York: Free Press, 1994).

Darwall, Stephen. "Grotius at the Creation of Modern Moral Philosophy." *Archiv fur Gestchichte der Philosophie* 94, no. 3 (October 2012).

Darwall, Stephen. "Hume and the Invention of Utilitarianism." In *Hume and Hume's Connexions,* ed. M. A. Stewart and John P. Wright (University Park: Penn State University Press, 1994).

Darwall, Stephen. "Norm and Normativity." In *The Cambridge History of Eighteenth-Century Philosophy*, ed. Knud Haakonssen (Cambridge: Cambridge University Press, 2006).

Darwin, Charles. *The Descent of Man*. In *The Works of Charles Darwin*, ed. P. H. Barrett and R. B. Freeman (London: Pickering and Chatto, 1989)

Darwin, Charles. *Notebook M: [Metaphysics on morals and speculations on expression (1838)]*, transcribed by Kees Rookmaaker, ed. Paul Barrett. *Darwin Online.* http://darwin-online.org.uk/.

Dear, Peter. *Revolutionizing the Sciences: European Knowledge and Its Ambitions, 1500–1700* (Princeton, NJ: Princeton University Press, 2001).

DeJean, Joan. *How Paris Became Paris* (New York: Bloomsbury, 2014).

de Laguna, Theodore. "Stages of the Discussion of Evolutionary Ethics." *Philosophical Review* 14, no. 5 (1905).

Dennett, Daniel. *From Bacteria to Bach and Back: The Evolution of Minds* (New York: Norton, 2017).

Dennett, Daniel. "The Self as a Center of Narrative Gravity." In *Self and Consciousness: Multiple Perspectives*, ed. F. Kessel, P. Cole, and D. Johnson (Hillsdale, NJ: Erlbaum, 1992).

DePaul, Michael. "Character Traits, Virtues, and Vices: Are There None?" In *Proceedings of the 20th World Congress of Philosophy*, vol. 1 (Bowling Green, OH: Philosophy Documentation Center, 1999).

Descartes, René. *Le Monde, ou Traite de la lumiere*. Trans. Michael Sean Mahoney (New York: Abaris Books, 1979).

de Waal, Frans B. M. *The Age of Empathy: Nature's Lessons for a Kinder Society* (New York, Harmony, 2009).

de Waal, Frans. *The Bonobo and the Atheist* (New York: W. W. Norton and Company, 2013).

de Waal, Frans. *Good Natured: The Origins of Right and Wrong in Humans and Other Animals* (Cambridge, MA: Harvard University Press, 1996).

de Waal, Frans. *The Primate Mind* (Cambridge, MA: Harvard University Press, 2012).

de Waal, Frans B. M. *Primates and Philosophers: How Morality Evolved* (Princeton, NJ: Princeton University Press, 2009).

de Waal, Frans B. M. "Putting the Altruism Back into Altruism: The Evolution of Empathy." *Annual Review of Psychology* 59 (2008).

Diderot, Denis. s.v. "Philosophe." In *Encyclopédie, ou dictionnaire raisonné des sciences, des arts et des métiers, etc.,* eds. Denis Diderot and Jean le Rond d'Alembert. ARTFL Encyclopédie Project (Autumn 2017 Edition), Robert Morrissey and Glenn Roe (eds.). http://encyclopedie.uchicago.edu/.

Dijksterhuis, Eduard Jan. *The Mechanization of the World Picture: Pythagoras to Newton.* (Princeton, NJ: Princeton University Press, 1986).

Donagan, Alan. "Twentieth-Century Anglo-American Ethics." In *A History of Western Ethics*, ed. Lawrence C. Becker and Charlotte B. Becker (New York: Routledge, 2003).

Doris, John, and Stephen Stich. "Moral Psychology: Empirical Approaches." In *The Stanford Encyclopedia of Philosophy* (Fall 2014 edition), ed. Edward N. Zalta. http://plato.stanford.edu.

Driver, Julia. "The History of Utilitarianism." In *The Stanford Encyclopedia of Philosophy* (Winter 2014 edition), ed. Edward N. Zalta. http://plato.stanford.edu.

Driver, Julia. "Moral Sense and Sentimentalism." *The Oxford Handbook of the History of Ethics,* ed. Roger Crisp (Oxford: Oxford University Press, 2013).

Duke, Aaron A., and Laurent Begue. "The Drunk Utilitarian: Blood Alcohol Concentration Predicts Utilitarian Responses in Moral Dilemmas." *Cognition* 134 (2015).

Farber, Paul Lawrence. *The Temptations of Evolutionary Ethics* (Berkeley: University of California Press, 1994).

Flanagan, Owen. *The Really Hard Problem* (Cambridge, MA: MIT Press, 2007).

Flanagan, Owen. "Varieties of Naturalism." In *The Oxford Handbook of Religion and Science*, ed. Philip Clayton and Zachary Simpson (Oxford: Oxford University Press, 2006).

Fodor, Jerry. *A Theory of Content and Other Essays* (Cambridge, MA, Bradford Books/MIT Press, 1990).

Fogelin, Robert. *Hume's Skeptical Crisis* (New York: Oxford University Press, 2009).

Foster, Kevin R., Tom Wenseleers, and Francis L. W. Ratnieks. "Kin Selection Is the Key to Altruism." *Trends in Ecology & Evolution* 21, no. 2 (February 2006).

Foucault, Michel. *The Birth of the Clinic: An Archaeology of Medical Perception* (London: Tavistock, 1973).

Foucault, Michel. *Power/Knowledge* (Brighton: Harvester, 1984).

Frankel, Charles. *The Faith of Reason: The Idea of Progress in the French Enlightenment* (New York: King's Crown Press, Columbia University, 1949).

Galilei, Galileo. *Opere Complete di Galileo Galilei.* (Florence, 1842).

Gay, Peter. *The Enlightenment: An Interpretation*, vol. I, *The Rise of Modern Paganism* (New York: Alfred. A. Knopf, 1966).

Gazzaniga, Michael S. *The Ethical Brain: The Science of Our Moral Dilemmas.* (New York: Harper-Perennial, 2006).

Gill, Michael B. "Ethics and Sentiment: Shaftesbury and Hutcheson." *The Routledge Companion to Ethics,* ed. John Skorupski (London: Routledge, 2010).

Goffman, Erving. *Frame Analysis: An Essay on the Organization of Experience* (Harmondsworth, England: Penguin, 1974).

Goffman, Erving. "The Interaction Order." *American Sociological Review* 48 (1983).

Graham, Jesse, Jonathan Haidt, Sena Koleva, Matt Motyl, Ravi Iyer, Sean P. Wojcik, and Peter H. Ditto. "Moral Foundations Theory: The Pragmatic Validity of Moral Pluralism." SSRN Scholarly Paper, Social Science Research Network (Rochester, New York: November 28, 2012).

Grant, Edward. *Foundations of Modern Science in the Middle Ages* (Cambridge: Cambridge University Press, 1996).

Green, T. H. *Prolegomena to Ethics*, ed. A. C. Bradley (Oxford: Oxford University Press, 1883).

Greene, Joshua D. "Beyond Point-and-Shoot Morality: Why Cognitive (Neuro) Science Matters for Ethics." *Ethics* 124, no. 4 (2014).

Greene, Joshua. *Moral Tribes: Emotion, Reason, and the Gap between Us and Them* (New York: Penguin Press, 2013).

Greene, Joshua D. "The Philosopher in the Theater." In *The Social Psychology of Morality: Exploring the Causes of Good and Evil*, ed. M. Mikulincer & P. R. Shaver (Washington, DC: American Psychological Association, 2012).

Greene, Joshua. "Social Neuroscience and the Soul's Last Stand." In *Social Neuroscience: Toward Understanding the Underpinnings of the Social Mind*, ed. A.

Todorov, S. Fiske, and D. Prentice (New York: Oxford University Press, 2011).

Greene, Joshua D. "The Terrible, Horrible, No-Good, Very Bad Truth about Morality and What to Do about It." PhD diss. (Princeton University, 2002).

Greene, Joshua D., Leigh E. Nystrom, Andrew D. Engell, John M. Darley, and Jonathan D. Cohen. "The Neural Bases of Cognitive Conflict and Control in Moral Judgment." *Neuron* 44, no. 2 (October 14, 2004).

Greene, Joshua D., R. Brian Sommerville, Leigh E. Nystrom, John M. Darley, and Jonathan D. Cohen. "An fMRI Investigation of Emotional Engagement in Moral Judgment." *Science* 293, no. 5537 (September 14, 2001).

Grotius, Hugo. *The Rights of War and Peace.* (repr., Indianapolis, IN: Liberty Fund, 2005).

Gutting, Gary. "Sam Harris's Vanishing Self." *New York Times*, September 7, 2014. https://opinionator.blogs.nytimes.com.

Haakonssen, Knud. "Early Modern Natural Law." In *The Routledge Companion to Ethics*, ed. John Skorupski (Oxford: Routledge, 2010).

Haakonssen, Knud. *Natural Law and Moral Philosophy* (Cambridge: Cambridge University Press, 1996).

Haakonssen, Knud. *The Science of a Legislator* (Cambridge: Cambridge University Press, 1981).

Haidt, Jonathan. "Morality." *Perspectives on Psychological Science* 3, no. 1 (2008).

Haidt, Jonathan. *The Righteous Mind: Why Good People Are Divided by Politics and Religion* (New York: Vintage Books, 2012).

Haidt, Jonathan, and Jesse Graham. "When Morality Opposes Justice: Conservatives Have Moral Intuitions That Liberals May Not Recognize." *Social Justice Research* 20, no. 1 (May 23, 2007).

Halevy, Elie. *The Growth of Philosophic Radicalism* (New York: Macmillan Company, 1928).

Hampton, Jean. "Hobbes and Ethical Naturalism." *Philosophical Perspectives* 6, Ethics (1992).

Harris, Sam. *The Moral Landscape: How Science Can Determine Human Values* (New York: Free Press, 2010).

Harrison, Peter. *The Territories of Science and Religion* (Chicago: The University of Chicago Press, 2015).

Hartly, David. *Observations on Man, His Frame, His Duty, and His Expectations* (London: Richardson, 1749).

Hartshorne, Hugh, and Mark A. May. *Studies in the Nature of Character*, vol. 1, *Studies in Deceit.* (New York: Macmillan, 1928).

Hattab, Helen. *Descartes on Forms and Mechanisms* (Cambridge: Cambridge University Press, 2009).

Hauser, Marc D. *Moral Minds: How Nature Designed Our Universal Sense of Right and Wrong* (London: Abacus, 2006).

Helvetius, Claude. *De L'esprit: Or, Essays On the Mind, And Its Several Faculties* (London: Printed for J. M. Richardson and Sherwood, Neely, and Jones, 1809).

Henrich, Joseph, Steven Heine, and Ara Norenzayan. "Most People Are Not WEIRD." *Nature* 466, no. 1 (2010).

Henrich, Joseph, Steven Heine, and Ara Norenzayan. "The Weirdest People in the World." *Behavioral and Brain Sciences* 33, nos. 2–3 (June 2010).

Hobbes, Thomas. *Leviathan.* (Oxford: Clarendon Press, 1909).

Hughes, William O. H., Benjamin P. Oldroyd, Madeleine Beekman, and Francis L. W. Ratnieks. "Ancestral Monogamy Shows Kin Selection Is Key to the Evolution of Eusociality." *Science* 320, no. 5880 (May 30, 2008).

Hume, David. "Letter to Frances Hutcheson." In John Hill Burton, *Life and Correspondence of David Hume* (Edinburgh: William Tait, 1846).

Hume, David. *A Treatise of Human Nature*, ed. David Fate Norton and Mary J. Norton (Oxford: Oxford University Press, 2000).

Hunter, Ian. *Rival Enlightenments* (Cambridge: Cambridge University Press, 2001).

Hunter, James Davison. *Culture Wars: The Struggle to Define America* (New York: Basic Books, 1991).

Hunter, James Davison, and Paul Nedelisky. "Where the New Science of Morality Goes Wrong." *The Hedgehog Review* 18, No. 3 (Fall 2016).

Hurka, Thomas. "Moore's Moral Philosophy." In *The Stanford Encyclopedia of Philosophy* (Summer 2010 edition), ed. Edward N. Zalta. http://plato.stanford.edu.

Huxley, T. H. "Evolution and Ethics." *Evolution and Ethics and Other Essays* (New York: D. Appleton and Company, 1905).

Iliffe, Rob. *Priest of Nature: The Religious Worlds of Isaac Newton* (New York: Oxford University Press, 2017).

Irwin, Terence. *The Development of Ethics*, 3 vols. (New York: Oxford University Press, 2008).

Isen, A. M., and P. F. Levin. "Effect of Feeling Good on Helping: Cookies and Kindness." *Journal of Personality and Social Psychology* 21 (1972).

Janiak, Andrew. "Newton's Philosophy." In *The Stanford Encyclopedia of Philosophy* (Summer 2014 edition), ed. Edward N. Zalta. http://plato.stanford.edu.

Johnson, Mark. *Morality for Humans* (Chicago: University of Chicago Press, 2014).

Johnson, Robert. *Rational Morality: A Science of Right and Wrong* (Norfolk, England: Dangerous Little Books, 2013).

Kahane, Guy. "Intuitive and Counterintuitive Morality." In *Moral Psychology and*

Human Agency: Philosophical Essays on the Science of Ethics, ed. Justin D'Arms and Daniel Jacobson (Corby, England: Oxford University Press, 2014).

Kahane, Guy, Katja Wiech, Nicholas Shackel, Miguel Farias, Julian Savulescu, and Irene Tracey. "The Neural Basis of Intuitive and Counterintuitive Moral Judgment." *Social Cognitive and Affective Neuroscience* 7, no. 4 (2012).

Kahneman, Daniel, and Alan B. Krueger. "Developments in the Measurement of Subjective Well-Being." *Journal of Economic Perspectives* 20, no. 1 (Winter 2006).

Kass, Leon R. "The Wisdom of Repugnance." *New Republic* 216, no. 22 (June 2, 1997).

Kellner, Hansfried, and Frank W. Heuberger. *Hidden Technocrats: The New Class and New Capitalism* (New Brunswick, NJ: Transaction Publishers, 1992).

Kemp Smith, Norman. *The Philosophy of David Hume* (London: Macmillan & Co., 1941).

Kitcher, Philip. *The Ethical Project* (Cambridge, MA: Harvard University Press, 2011).

Kitcher, Philip. "Naturalistic Ethics without Fallacies." *Preludes to Pragmatism: Toward a Reconstruction of Philosophy* (New York: Oxford University Press, 2012).

Kitcher, Philip. *Vaulting Ambition* (Cambridge, MA: MIT Press, 1987).

Koenigs, M., L. Young, R. Adolphs, D. Tranel, F. Cushman, M. Hauser, & A. Damasio. "Damage to the Prefrontal Cortex Increases Utilitarian Moral Judgments." *Nature* 446, no. 7138 (2007).

Kosfeld, Michael, Markus Heinrichs, Paul J. Zak, Urs Fischbacher, and Ernst Fehr. "Oxytocin Increases Trust in Humans." *Nature* 435 (June 2, 2005).

Kronqvist, Camilla. "The Relativity of Westermarck's Moral Relativism." In *Westermarck*, Occasional Paper No. 44 of the Royal Anthropological Institute, ed. David Shankland (Herefordshire, England: Sean Kingston Publishing, 2014).

Kupperman, J. J. "The Indispensability of Character." *Philosophy* 76 (2001).

Laclau, Ernesto, and Chantal Mouffe. *Hegemony and Socialist Strategy* (London: Verso, 1985).

Lears, Jackson. "Get Happy." *The Nation*, November 6, 2013.

Leiter, Brian. "Normativity for Naturalists." *Philosophical Issues* 25, no. 1 (2015).

Leroi, Armond Marie. "One Republic of Learning: Digitizing the Humanities." *New York Times*, February 13, 2015.

Levitin, Daniel J. *The Organized Mind: Thinking Straight in the Age of Information Overload* (New York: Penguin, 2014).

Locke, John. *An Essay Concerning Human Understanding,* Book II. http://www.earlymoderntexts.com/assets/pdfs/locke1690book2.pdf.

Locke, Don. "A Psychologist among the Philosophers: Philosophical Aspects of

Kohlberg's Theories." In *Lawrence Kohlberg: Consensus and Controversy*, ed. Sohan Modgil and Celia Modgil (London: Routledge, 2011).

LoLordo, Antonia. "Epicureanism and Early Modern Naturalism." *British Journal for the History of Philosophy* 19, no. 4 (2011).

Long, Douglas G. "Science and Secularization in Hume, Smith, and Bentham." In *Religion, Secularization, and Political Thought: Thomas Hobbes to J. S. Mill*, ed. James E. Crimmins (London: Routledge, 1990).

MacCulloch, Diarmaid. *All Things Made New: The Reformation and Its Legacy* (Oxford: Oxford University Press, 2016).

Machin, Alfred. *Darwin's Theory Applied to Mankind* (New York: Longmans, Green and Co., 1937).

Mackie, J. L. *Ethics: Inventing Right and Wrong* (London: Pelican Books, 1977).

Mathews, K. E., and L. K. Cannon. "Environmental Noise Level as a Determinant of Helping Behavior." *Journal of Personality and Social Psychology* 32 (1975).

McMahon, Darrin. "The Pursuit of Happiness in History." In *The Oxford Handbook of Happiness*, ed. Susan A. David, Ilona Boniwell, and Amanda Conley Ayers (Oxford: Oxford University Press, 2013).

Mill, John Stuart. "Bentham." In *Utilitarianism and On Liberty*, ed. M. Warnock, (Oxford: Wiley-Blackwell, 2008).

Mill, John Stuart. *Utilitarianism* (Indianapolis, IN: Hackett Publishing, 2001).

Miller, George A. "The Cognitive Revolution: A Historical Perspective." *TRENDS in Cognitive Sciences* 7, no. 3 (March 2003).

Mikulincer, M., and P. R. Shaver, eds. *The Social Psychology of Morality: Exploring the Causes of Good and Evil* (Washington, DC: American Psychological Association, 2012).

Mivart, St. George Jackson. *Genesis of Species* (New York: Appleton, 1871).

Modgil, Sohan, and Celia Modgil, eds. *Lawrence Kohlberg: Consensus and Controversy* (London: Routledge, 2011).

Montaigne, Michel de. *Essays* (Urbana, Illinois: Project Gutenberg, 2006). https://www.gutenberg.org.

Moore, G. E. *Principia Ethica* (Cambridge: Cambridge University Press, 1903).

Morris, Stephen G. *Science and the End of Ethics* (New York: Palgrave Macmillan, 2015).

Murphy, Mark. "The Natural Law Tradition in Ethics." In *The Stanford Encyclopedia of Philosophy* (Winter 2011 edition), ed. Edward N. Zalta. http://plato.stanford.edu.

Nadler, Stephen. "Doctrines of Explanation in Late Scholasticism and in the Mechanical Philosophy." In *The Cambridge History of Seventeenth-Century Philosophy*, ed. D. Garber and M. Ayers (Cambridge: Cambridge University Press, 1998).

Nagel, Thomas. "The Taste for Being Moral." *New York Review of Books*, December 6, 2012.

Nave, Gideon, Colin Camerer, and Michael McCullough. "Does Oxytocin Increase Trust in Humans? A Critical Review of Research." *Perspectives on Psychological Science* 10, no. 6 (November 2015).

Neimark, Neil. *The Science of Positive Thinking* (Irvine, CA: Author, 2015).

Nettle, Daniel. *Happiness: The Science behind Your Smile* (New York: Oxford University Press, 2006).

Noë, Alva. "Are the Mind and Life Natural?" National Public Radio, October 12, 2012.

Nussbaum, Martha. "Who Is the Happy Warrior? Philosophy Poses Questions to Psychology." *Journal of Legal Studies*, 37, S2 (June 2008).

Oettingen, Gabriele. *Rethinking Positive Thinking: Inside the New Science of Motivation* (New York: Penguin, 2015).

Ogden, C. K. *Bentham's Theory of Fictions* (New York: Harcourt Brace & Company, 1932).

Okasha, Samir. "Biological Altruism." In *The Stanford Encyclopedia of Philosophy* (Fall 2013 edition), ed. Edward N. Zalta. http://plato.stanford.edu.

Oliner, Samuel P., and Pearl M. Oliner. *The Altruistic Personality: Rescuers of Jews in Nazi Europe* (New York: Free Press, 1988).

Paley, William. *Moral and Political Philosophy*. In *The Works of William Paley* (London: T. Nelson & Sons, 1851).

Pawelski, James O. "Introduction to Philosophical Approaches to Happiness." In *The Oxford Handbook of Happiness*, ed. Susan A. David, Ilona Boniwell, and Amanda Conley Ayers (Oxford: Oxford University Press, 2013).

Pigliucci, Massimo, and Maarten Boudry, eds. *Philosophy of Pseudoscience: Reconsidering the Demarcation Problem* (Chicago: University of Chicago Press, 2013).

Pinker, Steven. *The Better Angels of Our Nature: Why Violence Has Declined,* reprint ed. (New York: Penguin Books, 2012).

Pinker, Steven. *Enlightenment Now: The Case for Reason, Science, Humanism, and Progress* (New York: Viking, 2018).

Pinker, Steven. "The Moral Instinct." *New York Times*, January 13, 2008.

Pinker, Steven. "Science Is Not Your Enemy." *New Republic*, August 6, 2013.

Porter, Roy. *The Enlightenment* (New York: Palgrave: 2001).

Preston, S. D., and Frans B. M. de Waal. "Empathy: Its Ultimate and Proximate Bases." *Behavioral and Brain Sciences* 25, no. 1 (2002).

Pufendorf, Samuel von. *On the Duty of Man and Citizen*, ed. James Tully (Cambridge: Cambridge University Press, 2003).

Randall, John Herman, Jr. *The Making of the Modern Mind: A Survey of the Intellectual Background of the Present Age* (Boston: Hoghton Mifflin Co., 1926).

Rawls, John. *A Theory of Justice* (Cambridge, MA: Belknap Press of Harvard University Press, 1971).

Richards, Robert J. *Darwin and the Emergence of Evolutionary Theories of Mind and Behavior* (Chicago: University of Chicago Press, 1987).

Robinson, Howard. "Dualism." In *The Stanford Encyclopedia of Philosophy* (Winter 2012 edition), ed. Edward N. Zalta. http://plato.stanford.edu.

Rogers, G. A. J. "Locke, Law, and the Law of Nature." In *The Empiricists: Critical Essays on Locke, Berkeley, and Hume,* ed. Margaret Atherton (Lanham, MD: Rowman & Littlefield Publishers, 1999).

Rose, Nikolas, and Joelle M. Abi-Rached. *Neuro: The New Brain Sciences and the Management of the Mind* (Princeton, NJ: Princeton University Press, 2013).

Rosen, Frederick. *Classical Utilitarianism from Hume to Mill* (London: Routledge, 2015).

Rosenberg, Alex. *The Atheist's Guide to Reality: Enjoying Life Without Illusions* (New York: W. W. Norton and Company, 2011).

Rosenberg, Alex. "Disenchanted Naturalism." In *Contemporary Philosophical Naturalism and Its Implications,* ed. Bana Bashour and Hans Muller (New York: Routledge, 2014).

Russell, Bertrand. *The Impact of Science on Society* (New York: Simon & Schuster, 1953).

Schneewind, J. B. *Sidgwick's Ethics and Victorian Moral Philosophy* (Oxford: Oxford University Press, 1977).

Schneewind, J. B. *The Invention of Autonomy* (Cambridge: Cambridge University Press, 1998).

Schuessler, Jennifer. "An Author Attracts Unlikely Allies." *New York Times*, February 7, 2013.

Seidler, Michael. "Pufendorf's Moral and Political Philosophy." In *The Stanford Encyclopedia of Philosophy* (Spring 2013 Edition), ed. Edward N. Zalta. http://plato.stanford.edu.

Seligman, Martin. *Flourish: A Visionary New Understanding of Happiness and Well-Being* (New York: Free Press, 2011).

Seppala, Emma. *The Happiness Track: How to Apply the Science of Happiness to Accelerate Your Success* (New York: HarperOne, 2016).

Shapin, Steven. *The Scientific Revolution* (Chicago: University of Chicago Press, 1996).

Shapin, Steven, and Simon Schaffer. *Leviathan and the Air-Pump: Hobbes, Boyle, and the Experimental Life* (Princeton, NJ: Princeton University Press, 1985).

Shaver, Robert. "Utilitarianism: Bentham and Rashdall." In *The Oxford Handbook of the History of Ethics* ed. Roger Crisp (Oxford: Oxford University Press, 2013).

Shaw, Tamsin, Steven Pinker, and Jonathan Haidt. "Moral Psychology: An Exchange." *New York Review of Books,* April 7, 2016.

Shen, Helen. "Neuroscience: The Hard Science of Oxytocin." *Nature* 522 (June 25, 2015).

Shermer, Michael. "Can Science Determine Moral Values? A Challenge from and Dialogue with Marc Hauser about The Moral Arc." *The Moral Arc,* http://moralarc.org.

Shermer, Michael. *The Moral Arc: How Science and Reason Lead Humanity toward Truth, Justice, and Freedom* (New York: Henry Holt, 2015).

Shermer, Michael. "The Science of Right and Wrong: Can Data Determine Moral Values?" *Scientific American,* January 1, 2011.

Sider, Theodore. *Four Dimensionalism: An Ontology of Persistence and Time* (New York: Oxford University Press, 2001).

Sidgwick, Henry. *The Methods of Ethics,* 7th ed. (London: Macmillan and Co., 1907).

Sidgwick, Henry. *Outlines of a History of Ethics,* 5th ed. (1902; repr. Indianapolis, IN: Hackett Publishing Company, 1988).

Sommers, Tamler. *A Very Bad Wizard: Morality behind the Curtain* (San Francisco: McSweeney's, 2009).

Sommers, Tamler, and Alex Rosenberg. "Darwin's Nihilistic Idea: Evolution and the Meaninglessness of Life." *Biology and Philosophy* 18, no. 5 (November 2003).

Sorley, William Ritchie. *The Ethics of Naturalism: A Criticism* (Edinburgh: William Blackwood and Sons, 1904).

Spencer, Herbert. *The Principles of Ethics,* vol. 1 (New York: D. Appleton and Co., 1892).

Sreenivasan, Gopal. "Errors about Errors: Virtue Theory and Trait Attribution." *Mind* 111 (January 2002).

Stephen, Leslie. *The Science of Ethics* (New York: G. P. Putnam's Sons, 1882).

Sturgeon, Nicholas. "Ethical Naturalism." In *The Oxford Handbook of Ethical Theory,* ed. David Copp (New York: Oxford University Press, 2006).

Tancredi, Laurence R. *Hardwired Behavior: What Neuroscience Reveals about Morality* (New York: Cambridge University Press, 2005).

Tanney, Julia. "Gilbert Ryle." In *The Stanford Encyclopedia of Philosophy* (Winter 2014 edition), ed. Edward N. Zalta. http://plato.stanford.edu.

Thagard, Paul. *The Brain and the Meaning of Life* (Princeton, NJ: Princeton University Press, 2010).

Thagard, Paul. "Eleven Dogmas of Analytic Philosophy." *Psychology Today,* December 12, 2014.

Thagard, Paul. "Nihilism, Skepticism, and Philosophical Method: A Response to Landau on Coherence and the Meaning of Life." *Philosophical Psychology* 26, no. 4 (2013).

Thompson, Paul. "Editor's Introduction." In *Issues in Evolutionary Ethics* (Albany: State University of New York Press, 1995).

Todorov, A., S. Fiske, and D. Prentice, eds. *Social Neuroscience: Toward Understanding the Underpinnings of the Social Mind* (New York: Oxford University Press, 2011).

Tooby, John, and Leda Cosmides. "Conceptual Foundations of Evolutionary Psychology." In *Handbook of Evolutionary Psychology*, ed. D. M. Buss (Hoboken, NJ: Wiley, 2005).

Trivers, Robert L. "The Evolution of Reciprocal Altruism." *Quarterly Review of Biology* 46, no. 1 (1971).

Tuck, Richard. "Grotius, Carneades, and Hobbes." *Grotiana* 4 (new series): 43–62 (1983).

Tuck, Richard. *Prolegomena to Hugo Grotius, The Rights of War and Peace* (repr. Indianapolis, IN: Liberty Fund, 2005).

Wallace, Alfred Russel. *Contributions to the Theory of Natural Selection* (New York: Macmillan and Co., 1870).

Watson, John B. *Behavior: An Introduction to Comparative Psychology* (New York: Henry Holt, 1914).

Watson, John B. "Psychology as the Behaviorist Views It." *Psychological Review* 20, no. 2 (1913).

Weber, Max. "Science as a Vocation." In *From Max Weber: Essays in Sociology*, trans. and ed. H. H. Gerth and C. Wright Mills (New York: Oxford University Press, 1946).

Wellmon, Chad. "Why Google Isn't Making Us Stupid . . . or Smart." *The Hedgehog Review* 14, no. 1 (Spring 2012).

Westermarck, Edward. *The Origin and Development of the Moral Ideas* (London: Macmillan and Co., 1906).

Westfall, Richard S. *The Construction of Modern Science: Mechanisms and Mechanics* (Cambridge: Cambridge University Press, 1971).

White, Matthew. "Selected Death Tolls for Wars, Massacres and Atrocities before the 20th Century." *Necrometrics*. http://necrometrics.com.

Wielenberg, Eric. "Review: Greene Joshua, Moral Tribes: Emotion, Reason, and the Gap between Us and Them." *Ethics* 124, no. 4 (July 2014).

Wilson, David Sloan. *Does Altruism Exist?* (New Haven, CT: Yale University Press, 2015).

Wilson, David Sloan, and Edward O. Wilson. "Rethinking the Theoretical Foundation of Sociobiology." *Quarterly Review of Biology* 82, no. 4 (December 2007).

Wilson, E. O. "The Biological Basis of Morality." *The Atlantic*, April 1998.

Wilson, E. O. *Sociobiology: The New Synthesis* (Cambridge, MA: Belknap Press of Harvard University Press, 1975).

Wolff, Robert Paul. "What I Have Been Reading." *The Philosopher's Stone*, September 1, 2013. http://robertpaulwolff.blogspot.com.

Wright, Robert. *The Moral Animal: The New Science of Evolutionary Psychology* (New York: Vintage Books, 1994).

Wurman, Richard. *Information Anxiety* (New York: Doubleday, 1989).

Wuthnow, Robert. *Communities of Discourse* (Cambridge, MA: Harvard University Press, 1989).

Wuthnow, Robert. *Communities of Discourse: Ideology and Social Structure in the Reformation, the Enlightenment, and European Socialism* (Cambridge. MA: Harvard University Press, 1993).

Yong, Ed. "One Molecule for Love, Morality, and Prosperity?" *Slate*, July 17, 2012. www.slate.com.

Yong, Ed. "The Weak Science behind the Wrongly Named Moral Molecule." *The Atlantic*, November 13, 2015.

Zak, Paul J. *The Moral Molecule: The Source of Love and Prosperity* (New York: Dutton, 2012).

Zimmerman, Dean. "The Privileged Present: Defending an 'A-Theory' of Time." In *Contemporary Debates in Metaphysics*, ed. Ted Sider, John Hawthorne, and Dean Zimmerman (Malden, MA: Blackwell, 2008).

Index